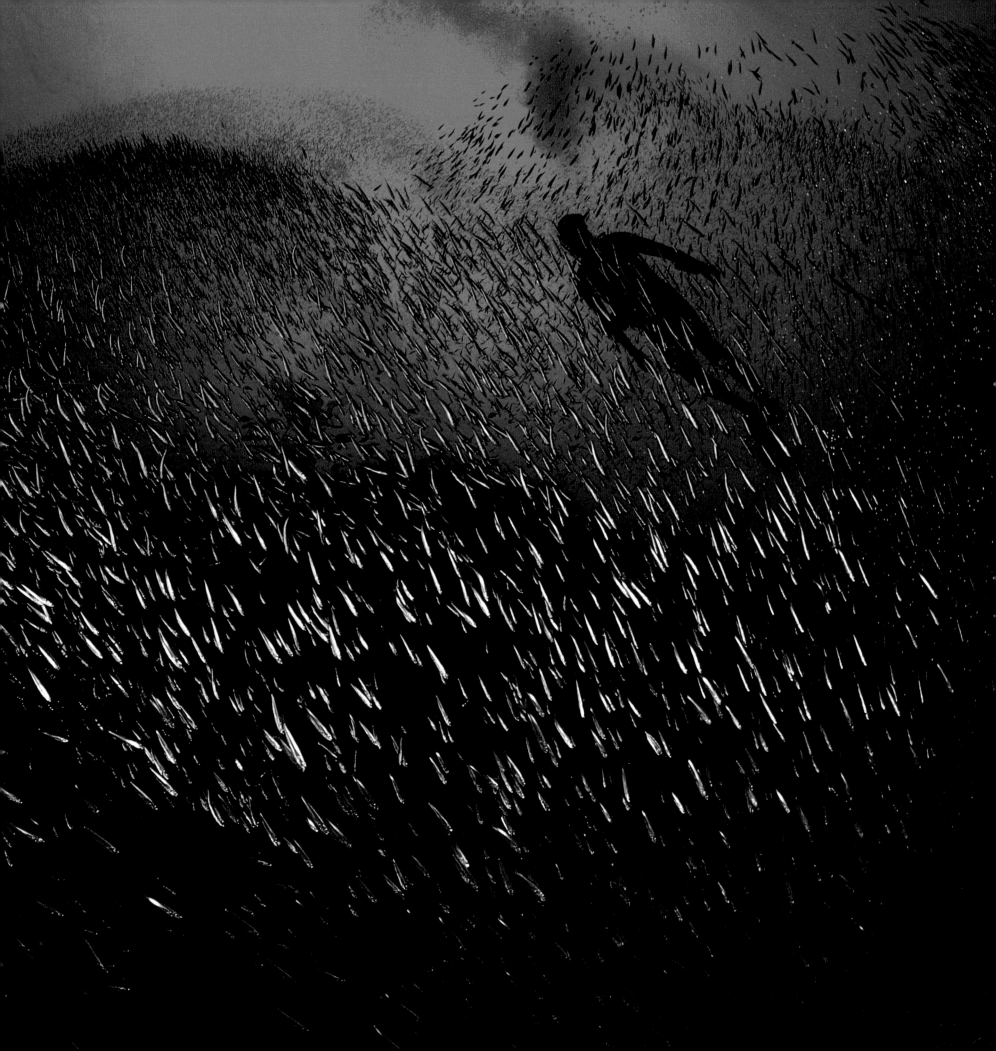

ACIFIC

AN UNDERSEA JOURNEY

DAVID DOUBILET

A BULFINCH PRESS BOOK

LITTLE, BROWN AND COMPANY

BOSTON · TORONTO · LONDON

ENDLEAVES: GREY SNAPPER SCHOOL, MARCHENA ISLAND, GALÁPAGOS

PAGE i: TOMATO CLOWNFISH IN BUBBLE-TIPPED ANEMONE, RABAUL, PAPUA NEW GUINEA

PAGES ii–iii: HARBOR SEAL SWIMMING THROUGH KELP, MONTEREY BAY, CALIFORNIA

PAGES iv–v: SCHOOLS OF SWEEPERS SWIRLING OVER THE "REEF AT THE EDGE OF THE WORLD," MILNE BAY, PAPUA NEW GUINEA

FRONTISPIECE: WHITE-FACED BLENNY, IZU PENINSULA, JAPAN

LIBRARY OF CONGRESS CATALOGING-IN-PUBLICATION INFORMATION APPEARS ON PAGE 191.

BULFINCH PRESS IS AN IMPRINT AND TRADEMARK OF LITTLE, BROWN AND COMPANY (INC.)
PUBLISHED SIMULTANEOUSLY IN CANADA BY LITTLE, BROWN & COMPANY (CANADA) LIMITED

PRINTED IN JAPAN

FOR MY FATHER, DR. HENRY DOUBILET

1906–1964

ASSOCIATE PROFESSOR OF SURGERY, NEW YORK UNIVERSITY

MAJOR, U.S. ARMY MEDICAL CORPS, RET.

YOU GAVE ME THE GIFT OF BOUNDLESS CURIOSITY.

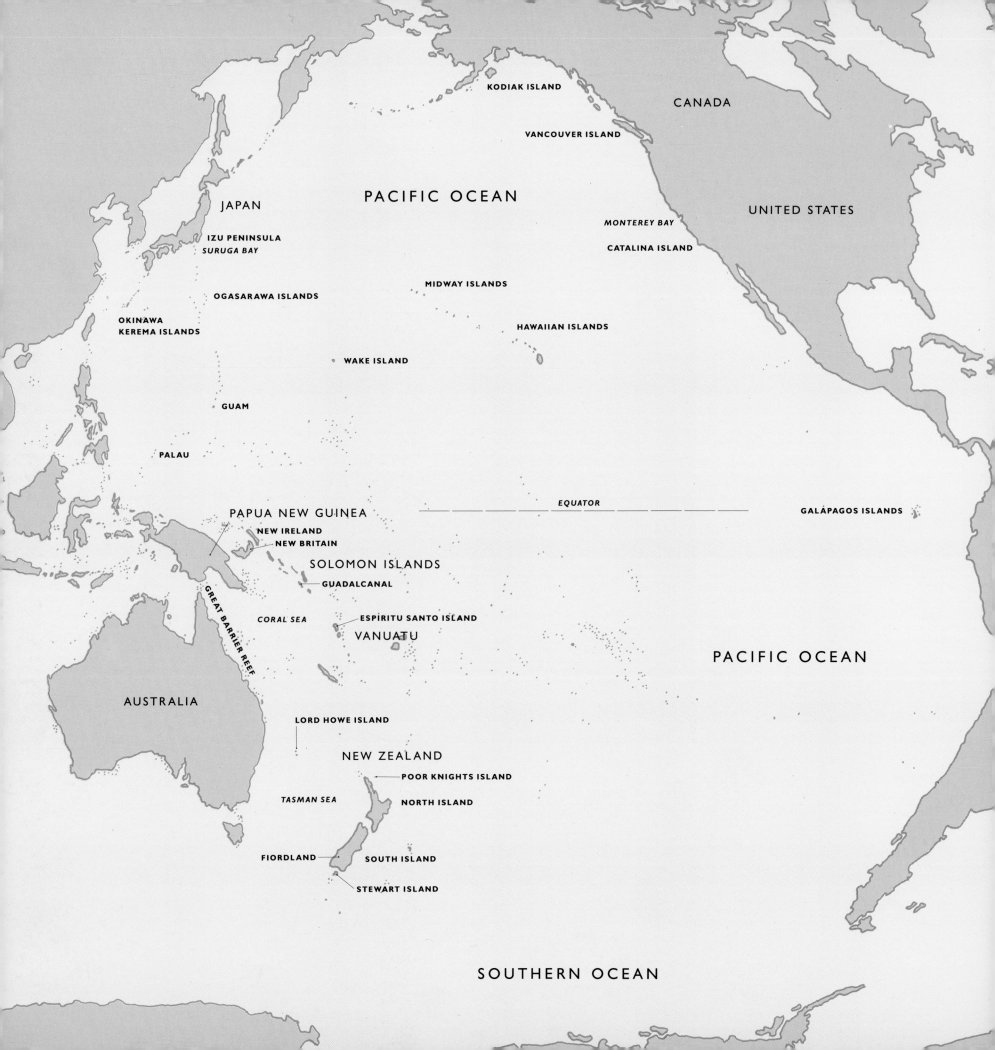

CONTENTS

INTRODUCTION

I was born on a coast where dawn comes from the sea, a bloom of silvery white summer light that spreads across the surface of the green-brown Atlantic Ocean. The bleak north New Jersey coast is punctuated by rock jetties that stick into the sea like dark, rotten teeth. In the summer the rich waters are a marine stew of plankton blooms and brown runoff from the land. On west wind days, the ocean near shore lies down. The water clears and the visibility leaps to an astounding ten feet.

When I was twelve, I began to hunt for striped bass. Swimming along the T jetties off Elberon, New Jersey, I kicked slowly through green fog, searching for a flick, a flash, of a silvertail. The stripers were like ghosts, fleeting, almost imaginary. Toward evening I would haul myself onto a jetty and watch the sunset over the land, beyond the breakwater, beyond the houses, gardens, and trees on the other side of Ocean Avenue. Finally, the sun would sink out of sight beyond the railroad tracks. The afterglow would be filled with the giant owllike moans of the Jersey Central and Pennsylvania trains.

I dreamed of a sea where I could see the sunset, a blue, clear sea.

I dreamed of the Pacific.

My aunt and uncle, Pauline and Johnny Falk, were, in a sense, Pacific explorers. In 1937 they were among the first paying passengers on Pan American's *China Clipper*. Pauline went down to the Chrysler Building to buy the tickets. In a tiny office she asked about the *China Clipper* and was handed a timetable, just like a railroad timetable. It read: San Francisco–Honolulu, Honolulu–Midway, Midway–Wake Island, Wake Island–Guam,

BOTTLE-NOSED DOLPHIN IN REFLECTED STORM CLOUDS, DOUBTFUL SOUND, FIORDLAND, SOUTH ISLAND, NEW ZEALAND

Guam–Manila, with times. And that was it. At Alameda in San Francisco Bay, they boarded the great whalelike flying boat. The plane, its four huge radial engines roaring, splashed westward down the bay, then lifted off, flying over the Golden Gate Bridge into the evening. It was a jump into the Pacific.

In 1971, I made such a jump. It was, however, easier, less fraught with the danger, dumped fuel, blown cylinders, and navigation that in 1937 was like crossing a trackless ocean on a tightrope. Dr. Ken Read, a professor of mine at Boston University, was taking a sabbatical with his family in Palau (now the Republic of Belau) in the Western Caroline Islands, seven hundred miles southwest of Guam. He invited me to come out while he studied the amazingly rich reefs of the western equatorial Pacific.

I flew on a Pan American plane, a Boeing 707 slightly old before its time. The plane's name was *Nightingale,* or more formally *The Clipper Nightingale. Nightingale* flew the cheapest and longest route across the Pacific. On a winter day, we went from New York to Fairbanks, Alaska. In the terminal there was an immense stuffed Kodiak bear with paws the size of my chest in a glass case.

We flew on. I sat in the last seat. The flight was full, stuffed like a sausage, mostly with military wives on their way to the Far East. We flew for almost ten hours, following the great circle route that arcs northward up and over the Aleutian Islands then slides southward across the shoulder of the globe to Tokyo. I looked out the window and saw only clouds. I couldn't sleep.

We landed in the rain at Haneda Airport in Tokyo. The transit lounge was a lagoon of glaring chrome and white waxed linoleum. I bought a pearl for my wife, Anne, in the duty-free shop, then sat down to watch *Gunsmoke* on television. Matt Dillon burst through the swinging doors of the Long Branch Saloon and rumbled in Japanese, *"Oh hio gozimas, Kitty."*

The plane flew on to Osaka, then finally, six hours away, Guam, which smelled of coconuts and jet fuel. There I changed planes to an Air Micronesia Lockheed Electra, not the kind that Amelia Earhart and Fred Noonan disappeared in, but the kind that once flew the Eastern Shuttle run. The first half of the plane was an enormous cargo bay, then came a place for life rafts, and finally the passenger compartment. This was no Eastern Shuttle. We left Guam on the way to Yap Island an hour before dawn, and finally, finally, I slept.

I awoke suddenly in Yap. Tropical sun was blasting through my window. A man in a loincloth with a bone in his nose was refilling the wing tanks. I was in the Pacific, the far reaches of the central Pacific, in Micronesia.

The next day I saw my first clownfish, a creature of inordinate beauty that does not inhabit the conservatively colored Caribbean. The clownfish flew over, then burrowed into, the yellow waving tentacles of an anemone. The anemone was perched at the edge of a drop-off—a vertical reef that fell into the blue dark depths. Two jet black manta rays swept along the edge of the wall, flying in the noonday light rays.

The Pacific Ocean covers nearly half of our planet, a great blue belly of our world. It is an ocean so vast that it makes islands out of continents. The southwest Pacific is the heart of the planet's coral seas. Somewhere in the complicated geography of Indonesia, Malaysia, and the Philippines, in the blood-warm water, lives a superrich coral environment. Here are reefs that have the most diverse sea life in the world. In this almost mythical corner of the Pacific, the coral life spreads out, out across a great belt of warm water that stretches westward from Easter Island in the east across the Indian Ocean to the very end of the Red Sea. Northward, the coral universe, this sixth continent made of tiny animals, touches the home islands of Japan. To the south, coral whispers at the edge of Lord Howe Island, the most southerly reef in the world.

But the Pacific is not a coral sea alone. It is an ocean ringed with cold water, with kelp forests and sea lions, squid and jellyfish. In New Zealand, where the waters of the South Pacific begin to mingle with the rich cold Southern Ocean, yellow-eyed penguins fly over dark sea bottoms and southern pigfish roll in surf that sweeps over sandy seabeds.

DR. WILLIAM HAMNER SURROUNDED BY MASTIGIAS JELLYFISH, JELLYFISH SALT LAKE,

EIL MALK ISLAND, PALAU, MICRONESIA

The Pacific Ocean is not a seamless, endless stretch of water. There are countries in the sea itself, countries whose borders are not defined by rivers or mountains. In the sea the borders are invisible barriers of distance and temperature, and great swaths of open ocean form deserts. These countries in the sea are inhabited by creatures that seem to share only the weightlessness of water.

This book is a collection of journeys across, around, and above the Pacific. They are journeys on assignment for *National Geographic*, so there was always a direct purpose. An assignment is strange business. It takes an absolute mountain of camera equipment to make a picture of a very tiny sea creature. The single most fearsome situation in underwater photography is not encountering man-eating sharks, giant, hideous octopuses, deadly stinging jellyfish, murderous ocean currents, or black bottomless depths, it is dealing with overweight, ridiculously excess baggage. There are reasons you cannot, no matter how fast you are, change film or lenses underwater. Everything must be lit. The sea, after all, "is a dark and evil place, full of gliding slimy creatures," as the master of arms in Herman Melville's *Billy Budd* said. I usually wind up with at least twenty bags of cameras, diving gear, and electronic flashes, all to photograph an inch-long goby. When the airline clerks see me coming they rub their hands together with glee.

There are rhythms in the oceans of our world. The distances between the peaks of waves are like an electrocardiogram that measures the heartbeat of an ocean. The Atlantic's rhythm is short and sharp, a gray and vicious rhythm. The Pacific's is a long slow breath, a slow heartbeat.

When I was ten my father took me bluefishing off Montauk Point on Long Island. The wind from the north was very strong, the sea glittering and corrugated. The fishing boat rolled angrily, corkscrewing and snapping between the valleys and peaks of the waves. The bluefish were out in force. I hooked one and fought it up to the boat, where a shaft of morning sunlight reflected off its cold fish eye. For some reason this sight released everything in my stomach and I said farewell to my "hardy fisherman's breakfast." The twisting, short-sharp waves of the Atlantic did me in.

The Pacific doesn't pummel, it squeezes. Anne and I once took a thirty-hour boat trip from Tokyo to the Ogasarawa Islands, part of a chain of volcanic islands ending at Iwo Jima. Amazingly, they are part of greater metropolitan Tokyo.

Except for an old World War II fighter strip there are no airfields, so the only way to go is the mini ocean liner, *Ogasarawa-Maru*. In 1980, she was a brand-new ship and very clean. There were, however, many classes. Anne and I traveled in something called supreme first class. I had convinced the *Geographic* travel office that I needed space to store all the equipment. The equipment, of course, ended up in the ship's hold, and we had a lovely cabin for ourselves. Supreme first class had its own bathroom and a small Japanese table with two legless chairs. Regular first class had four beds per room, with a shared bathroom down the hall; lower first class, six bunks and a communal bathroom; second class, a huge dormitory room with blue mats; and third class, an even bigger room—no mats.

So we sat in our ultrasupreme room at our low Japanese table and shared a giant peach, the size of a small grapefruit. We cruised down Tokyo Bay and by late afternoon, *Ogasarawa-Maru* turned the corner into the open Pacific and began to slide up and over the long swell. In the evening we went on deck. *Ogasarawa-Maru* was about two hundred feet too short for the long Pacific swells. Water hissed along the ship's flanks and the ship took on the rhythm of a tranquilized rocking horse. The ship was flooded with golden light. The water was very blue, almost black; we were crossing the Kuroshio—the black current, Japan's Gulf Stream.

The bow was pointed toward the Ogasarawa Islands, a place where we would waltz with sand tiger sharks and swim with a thousand bigeye fish that changed color, from silver to red, as they flowed through a clear, shallow coral lagoon. The sun touched a hazy horizon. I was in a sea where the sun sets. I was in the Pacific.

CLOWNFISH IN PURPLE-TIPPED ANEMONE, DINAH'S LAND, MILNE BAY, PAPUA NEW GUINEA

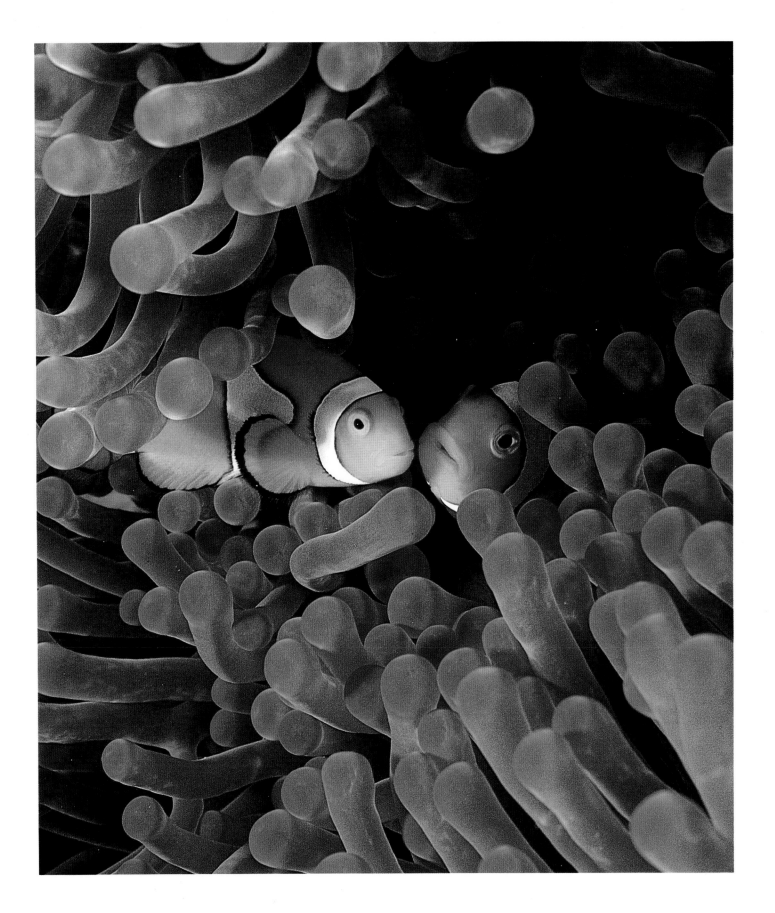

THE NORTH PACIFIC COAST

MONTEREY BAY

We went westward from Monterey across a pewter sea, beneath a dark gray sky. Sea and sky blended together. The air was cold and heavy with the leading edge of a winter storm. Thin, elegant, black-and-white northern right whale dolphins rose from the depths and momentarily rode the bow wave of our dive boat, the *Silver Prince*.

We had come fifteen miles out to the edge of Monterey Bay to watch the research ship *Atlantis II* launch the deep-diving submersible *Alvin*. Somewhere below, my colleague, *Geographic* photographer Emory Kristof, was cocooned in the steel sphere of *Alvin*, taking a midnight elevator ride to the edge of the Monterey Canyon. The great submarine canyon lay nearly four thousand feet below the keel of the *Silver Prince*. As complicated and convoluted as the Grand Canyon, the Monterey Canyon forms a kind of palatial staircase that leads from the shallows of Monterey Bay to the Pacific abyssal plain.

The dark sea was alive. Pods of Riso's dolphins with their blunt rounded heads crossed in front of the bow as large gatherings of California sea lions, two hundred or more in each group, hung like children, treading water, barking, splashing, and lying on their sides. The sea lions were migrating, swimming across the mouth of the bay, out of sight of land.

I wanted to get in with them, but it was a difficult situation. Usually if you stop a boat in midocean the marine mammals near the boat will vanish into the open sea. I looked at my diving partner, Jay Ireland. He shrugged and said, "Let's give it a try." The captain cut the engines of the *Silver Prince* and we slowly drifted into a group of barking, splashing sea lions. Jay and I put on our dry suits, masks, and flippers, then quietly slid over the side.

KELP FOREST, POINT LOBOS, CARMEL, CALIFORNIA

Trying to move gracefully in a baggy dry suit is like swimming in a tent, but we managed to insinuate ourselves into the sea lion herd. It was amazing; for every sea lion above water there seemed to be ten underneath. Diving and twisting, they barked and streamed bubbles like torpedoes in a ballet of anarchy. I gripped the camera housing tightly. In four thousand feet of water, a dropped camera is a lost camera. The open ocean was dark and menacing. When the *Silver Prince* had drifted 150 yards away, vanishing and reappearing in surface fog, I turned to Jay and said, "I have three words to say to you."

"Yes?"

I said, "Great. White. Shark."

We whirled around, searching the open sea. There *was* no shark, though this was the perfect place, a traveling buffet of sea lions that couldn't escape to land. The sea lions eventually got bored with us and moved on, barking and splashing across a gray seascape.

Monterey Bay is a meeting place between the southern California subtropical zone and northern temperate waters. From the spring through the fall, northwest winds help deflect the cold south-flowing California current. The deep canyon acts as a funnel, filling the bay with deep, rich water. The bay thrives on this cold oceanic stew. At times the surface boils with life, a street scene in the sea.

Just offshore, Jay and I watched as a sea lion got hold of a mola mola, or oceanic sunfish—a flat, tailless creature that can grow to weigh a thousand pounds. The sea lion bit its flippers off then flipped it in the air like a frisbee. These fish have a leatherlike armored skin and unappetizing flesh, so the sea lions don't eat them, they just torment them. By the time we could get into the water with the sea lion and the sunfish there would only be a half a mola mola left. Just plain mola.

We dove in the kelp beds off Point Pinos, near Pacific Grove at the southern edge of the bay. Kelp, the largest form of algae in the world, grows in giant strands from a holdfast base on the bottom to the surface more than one hundred feet above, forming a rain forest in the sea, where weightless divers can fly through the treetops. The surface sunlight filtered through the leaves, giving a slight golden glow to the sea. The fall water was clear and the kelp strands formed towering columns like those of a Greek temple. I found a rock completely coated with little cup anemones, feeding in the rich water. In the afternoon a harbor seal joined us. It swam through the kelp, twisting and turning. Pausing for a brief moment, the seal rested its head in the crook of a kelp frond.

At Point Lobos I watched a school of kelp bass swim beneath the canopy of leaves, just below the surface. A sun jellyfish drifted into the kelp forest as a small jack swam around its base. When I approached, the jack disappeared into the folds of the yolk-colored bell. The forest swayed and moved with the Pacific surge, but the kelp strands broke up the power of the waves.

South of Monterey the kelp forests become even more dramatic. Protected from the open Pacific by the bastions of southern California's Channel Islands, the kelp grows enormously tall in the clear open-ocean water. In the kelp forests off San Clemente Island I found a small cave with a large green moray eel in it. The walls of the cave were festooned with red shrimp. The shrimp would climb onto the eel's body and with their tiny pinchers pick bits of detritus off its soft, brown-velvet skin.

There is a particular stillness within the kelp forest, the stillness of a pine grove—a sanctuary in the enormous ocean.

PORTRAIT OF A HARBOR SEAL, MONTEREY BAY, CALIFORNIA

WALL OF SEA ANEMONES IN THE
HEART OF A KELP FOREST,
MONTEREY BAY, CALIFORNIA

15

LEFT:
HORN SHARK, CATALINA ISLAND,
CALIFORNIA

BELOW:
CABEZON FISH WATCHING AS
A STARFISH COVERS ITS EGGS,
MONTEREY BAY, CALIFORNIA

RIGHT:
A RED ROCK SHRIMP "CLEANING" THE
SKIN OF A CALIFORNIA MORAY EEL,
MONTEREY BAY, CALIFORNIA

JUVENILE PACIFIC POMPANO
SWIMMING IN THE TENTACLES OF
A SUN JELLYFISH, MONTEREY BAY,
CALIFORNIA

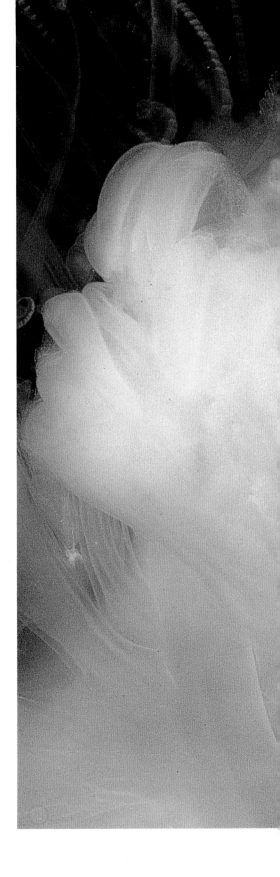

PACIFIC POMPANO HIDING IN
THE MANTLE OF A SUN JELLYFISH,
MONTEREY BAY, CALIFORNIA

SUN JELLYFISH UNDERNEATH A KELP CANOPY, MONTEREY BAY, CALIFORNIA

BENEATH THE KELP FOREST CANOPY, POINT LOBOS, CARMEL, CALIFORNIA

CALIFORNIA SEA LIONS MIGRATING
ACROSS MONTEREY BAY, CALIFORNIA

VANCOUVER ISLAND

As the west coast of the United States runs northward, the undersea countries at the edge of the Pacific begin to change. Life in the north Pacific grows larger, richer, stranger. The passage between mainland British Columbia and Vancouver Island is an emerald Eden of Pacific life. The Pacific pours through the Strait of Georgia, creating some of the most stupendous currents in the world. Off the town of Campbell River, the current screams past. My diving partner, Warren Buck, and I waited for slack water. Diving from a small cove that offered some protection from the current, we poked our face masks out into the current; bit by bit, it slackened and stopped. We swam out into a strange, dangerous land carpeted by strawberry anemones, an explosion of color and life. The current here bathed the seabed with food. Red Irish-lord rockfish blended with the bottom as hermit crabs scuttled across the tapestrylike carpet. I felt my flippers move, almost imperceptibly. The tide had changed. The current began as a first breath of wind before the storm. I hesitated —one more photograph—and then the still sea became a river. I dragged myself, hand over hand, to our little entry cove. Warren was waiting for me. He grabbed the back of my tank and dragged me to safety. My left flipper stuck out into the Strait and the current caught it and blew it off my foot. Annoyed, I watched it spin away, knowing that dive shops rarely sell just one flipper.

There is a place out of the roaring current at the edge of the Strait of Georgia. Sanitch Inlet, near the town of Sidney, is a steep-sided piece of water where the branches of pine trees overhang the bank. We dove here on a winter day. At 120 feet we found the cloud sponges, ancient creatures that live only in twilight. A ghostly white-yellow, they are like enormous swiss cheeses, full of holes and passages that form individual apartments for fish and crabs. I looked up the steep slope and watched my diving partners, bulky spacemen

PENPOINT GUNNEL ENTWINED IN RED SEA URCHIN, STRAIT OF GEORGIA, BRITISH COLUMBIA

in dry suits falling out of a green sky. A large twenty-pound octopus meandered down the submarine cliff. The Strait of Georgia is home to the largest octopuses in the world; weighing up to almost a hundred pounds, these creatures have no bones and seem able to stretch to infinite lengths, like Plastic Man. I waited, silently, not breathing, beside the cloud sponge. The octopus reached out with a delicate tentative gesture to touch the faceplate of my mask with a single tentacle. Withdrawing, it continued its liquid walk in the emerald dark of this corner of the Pacific.

KING CRAB BOAT, KODIAK ISLAND,
ALASKA

26

CLOWN SHRIMP ON PINK ANEMONE, STRAIT OF GEORGIA, BRITISH COLUMBIA

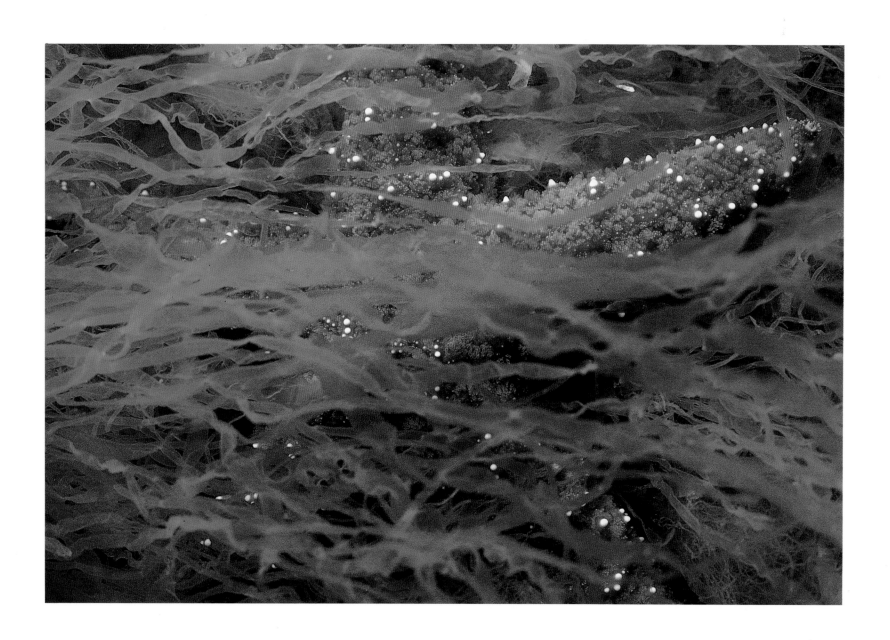

PURPLE STARFISH IN ALGAE, STRAIT OF GEORGIA, BRITISH COLUMBIA

PACIFIC PINK SCALLOP, VICTORIA, BRITISH COLUMBIA

PORTRAIT OF A WOLF EEL, RACE ROCKS, VANCOUVER, BRITISH COLUMBIA

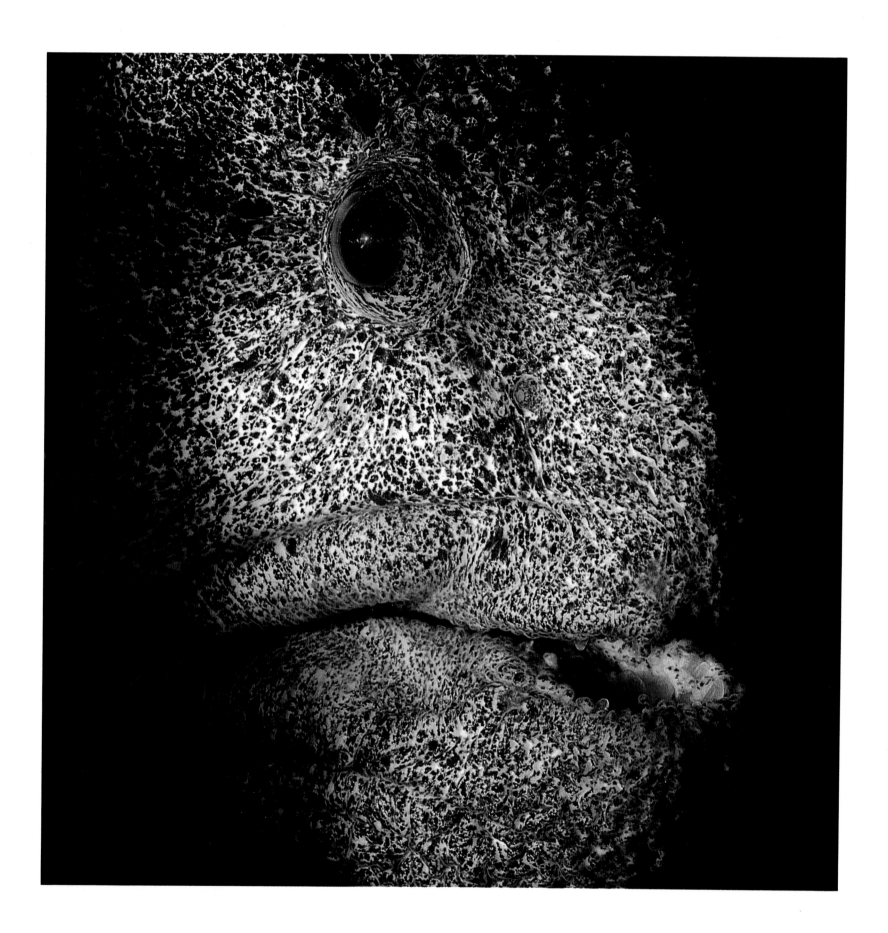

HERMIT CRABS, STRAIT OF GEORGIA, BRITISH COLUMBIA

GIANT OCTOPUS, VICTORIA, BRITISH COLUMBIA

ISLANDS

HAWAII

It is difficult to realize that Hawaii, the land of Waikiki, Don Ho, white shoes, volcanos, hotels full of tourists, giant waves, giant Hawaiians, and really ugly shirts, is the most isolated land on the face of our planet. It is a drop in the Pacific Ocean. Nothing is near the Hawaiian Islands. Great storms born in the western Pacific ride with the Kuroshio, the Japan Current, across the top of the Pacific, tuck into the Gulf of Alaska, and generate monstrous waves that march south until they touch the north coast of Oahu.

The Hawaiian Islands, with the Leeward Islands, including Midway, are distant out-riders of the Pacific coral community. Their isolation and colder waters do not support rich coral growth. The plate corals and reef-building stony corals here barely hold on to the volcanic bottom. The water is not the rich blue of the South Pacific. There is a touch of gray, an echo of the colder North Pacific. Much of the sea life is endemic to Hawaii; its colors are rich and vibrant, standing out against a darker, simpler seascape.

The islands are great volcanic mountains rising out of the deep Pacific abyssal plain, and the shallow waters are themselves mountainous, with sheer cliffs and vast caverns. Off Lanai, I swam into a cave that was a cathedral. The entrances were two soaring Gothic arches, full of afternoon sunlight. A school of yellow milletseed butterflyfish swam from darkness to light at the rim of the cave, a shelter in the open Pacific.

The seabed between the islands of Maui and Lanai is like a place of rolling hills, dotted with clumps of evergreens, a strange underseascape with a surface sky 220 feet above. The water is clear and the volcanic bottom seems to be bathed in strong blue moonlight. The evergreens are, in reality, black coral. This deep bottomland forms the richest black coral beds in the world. In 1978 I came to this place with Dr. Richard Grigg, legendary big wave rider, ex-aquanaut, professor of marine biology at the University of Hawaii, and coral expert. We were doing a story for *Geographic* on Hawaii's precious coral. The famous angel-skin pink coral was harvested by a small submarine from over a thousand feet down, but the black coral was taken by divers from the channel between Maui and Lanai. To reach the black coral beds is a fearsome, dangerous experience. Two hundred feet is beyond the safe limit for compressed-air diving, in the realm of nitrogen narcosis. The brain skates at the edge of blackout and vertigo here, and the lungs do not get enough nourishment from the impossibly thick compressed air. A great weight of water presses down from the distant surface. Overexertion can make any small mistake fatal, the ultimate form of drowning. The coral divers commute to this place every day. Ricky told me that the dive was easy, but he is an intellectual daredevil.

The divers found their secret dive site with the boat's fathometer, then took bearings from the green slopes of Maui. The boat driver ran upcurrent from the bearings and dropped us into the open sea. It is impossible to anchor in the three-knot currents that sweep through the channel and feed the coral beds. Ricky and I flew backward off the boat into open ocean, falling toward an invisible bottom, kicking and clearing our ears. The coral divers carried ten-pound sledge hammers, riding them to the bottom like anchors.

The trick to avoiding much of the effect of nitrogen narcosis is to reach the bottom at a high speed, breathing as little as possible. At eighty feet I saw the distant bottom. The other coral divers were tiny sticklike figures creeping toward the dark growths of coral trees, puffing endless streams of silver bubbles.

I hit the bottom like a parachutist, rolling and tumbling in the current. Ricky and I swam with the divers, watching them bang off the coral trees with their hammers. In the blue light the coral looked like Christmas trees. The divers tied the trees to a lift bag;

BREAKING WAVE, NORTH SHORE, OAHU, HAWAII

putting it over their heads, they exhaled into it, sending it up like a balloon rising through a watery sky. Each diver carried three lift bags, riding up to the surface on their last one. We followed, then decompressed hanging beneath a lift bag for an hour. A standard procedure; yet we were trapped in the open sea, slowly passing off nitrogen to avoid the bends. The blue moonlit land was invisible below. The empty Hawaiian sea was full of booming moans, the song of humpback whales that winter in the channel between Maui and Lanai.

STAR *II* SUBMERSIBLE RETURNING
FROM THE DEPTHS WITH A LOAD
OF PRECIOUS CORAL,
MOLOKAI CHANNEL, HAWAII

STAR *II* DIVING ALONG THE FLANK OF
THE MOLOKINI CRATER, MAUI, HAWAII

RETICULATED HAWKFISH, MAUI, HAWAII

RED-SPOTTED HAWKFISH, MAUI, HAWAII

BOTTLE-NOSED DOLPHIN AND CLOUDS,
KANEHOE BAY, HAWAII

BLUE TRIGGERFISH, MAUI, HAWAII

ORANGE SPOT SURGEON FISH, MOLOKINI CRATER, MAUI, HAWAII

CATHEDRAL CAVE, LANAI, HAWAII

THE GALÁPAGOS

In 1977 Anne and I visited the Galápagos Islands. This is what we took: ten diving tanks, 150 pounds of lead weights, two high-pressure compressors in two crates with compressor oil and charcoal filters, two forty-horsepower outboard motors with gas tanks and spare hoses, one Zodiak sixteen-foot inflatable boat, masks, fins, snorkles, wet suits, safety vests, five underwater camera housings, ten cameras, fifteen strobes, 120 pounds of strobe batteries, twenty-two lenses, three boxes of lens tissue, 450 rolls of film, two tripods, one tool box, one tape recorder with twenty-eight tapes, two bags of clothes, coffee, tea, and two bottles of Heinz catsup. Total: one and a half tons of overweight bags.

The trip required a year of planning. Gerry Wellington from the University of California at Santa Barbara came as a guide and writer. My friend and fellow photographer Michael O'Neill joined us, and Gil Grosvenor, then editor of *National Geographic* magazine, was also there. It was a grand adventure, almost an expedition, to photograph the sunken landscape of the Galápagos. This extraordinary part of the world, bound by currents six hundred miles off the coast of Ecuador, is an enchanted place full of ancient life.

But first we had to get there with the mountain of equipment. After five days, by some miracle, it all arrived. While we assembled the equipment we stayed at the Darwin Research Station on Santa Cruz Island. Anne and I shared a small dorm room with seventeen brown, hairy, butterplate-sized spiders that hung on the ceiling over the two steel-framed beds. A single suspended light bulb swayed back and forth, casting sinister shadows that magnified the horror of the spiders. I asked Gerry, wise Dr. Gerry, about how we should deal with the creatures. He said, "Don't touch them. They eat all the other insects. They are good for the balance of nature."

A wonderful response. He didn't have any spiders in his room to drop down while he was asleep. I asked the assistant director of the station what exactly we should do. He said, "I hate spiders. A lot." He proceeded to remove our spider population with a rolled-up 1975 issue of *Vogue*. It was a great event full of coordination, energy, and speed.

Finally, we went to sea in the station's research ship, *Beagle III,* the first of three boats I would charter in the next few months. We anchored in a cove full of sea lions, a nursery with hundreds of pups. The cameras and strobes were prepared, the rubber boat was inflated, and my new compressors were humming. I had brought two in case one broke—so smart. All was right with the world, and we were at the end of the world, straddling the equator in an almost mythical archipelago. Then both compressors quit, the gasoline engines dead, frozen. Captain Fiddi Angermeyer sniffed the gasoline and said with resignation, "Oh, no. They have done it again. They put gas into the molasses barrels. Look, smell it."

We opened up the compressors' Briggs and Stratton engines. Inside, they were full of crystallized molasses—a disgusting pink candy-coating on the valves, the pistons, the cylinders, the carburetors, everything. It took us hours of cleaning and chipping and then refiltering one hundred gallons of gasoline. It was well after midnight by the time the compressors were working, but sleep was out of the question. Hundreds of sea lions returned to the colony after gorging themselves on tons of squid in the deep night waters surrounding the islands. They proceeded to fight, bellow, bark, and barf, throwing up half-digested squid beaks from their huge feast. The wind shifted, enshrouding *Beagle III* with a cloud of odeur de sea lion.

Days later, we dove on a small, isolated rock outcropping that was home to another seal colony. It was a perfect morning, with a flat sea punctuated by strong equatorial sunlight. The sea lions spun down from a silver surface, playing, dancing, dive-bombing us. I shot quickly, then put my camera down on a rocky ledge and picked up another. Suddenly, prompted by a sixth sense, I looked back at the camera. Two sea lion pups were playing tug-o-war with it. I yelled, wagging my finger at them. They looked up with innocent, completely guilty, dog faces, then, with a twist and a turn, blasted off. The larger one, as he twisted, swam through the neck strap of the camera, which hung on him as if on a tourist. Inwardly I screamed, "No, no. It's my camera, not the office's—it's brand new. It's full of great pictures." All this complicated emotion came through my regulator as a series of grunts and bubbles. Fortunately, the brainless, giddy creature later dropped the camera. There were some wonderful pictures on that roll—I still wonder if the sea lion took them.

Underwater, the Galápagos are full of dark cliffs. The sea can be cold and green or warm and blue depending on the swirling currents which entwine the islands. In a few places coral, barely holding on in the cool waters, coats the sides of the islands. In Devil's Crown, a rock formation off Floreana, the little lagoon floor is carpeted by corals. A vast school of striped grunts filled up its bowl, forming a living false sky over our heads.

In the months we were in the Galápagos, we crossed the equator a dozen times, commuting between warm and cold oceans. Punta Espinosa on the island of Fernandina— also known as Narborough Island—is a place out of time, home to a huge group of marine iguanas. These creatures, the only seagoing lizards on the planet, cover a series of volcanic boulders at the edge of a cold, algae-rich green sea. They go into the sea to feed on the algae. It's wonderful to see them underwater—rocks, fish, seaweed, and an occasional sea lion, and then a minidragon. They clamp to the rock with their claws and graze on algae, rolling their little dragon heads back and forth. The sea lions bully and incessantly torment the humorless iguanas, pulling their tails and flipping them around, but they do not eat them. Maybe they think that they are too ugly and taste funny.

I swam up to a feeding iguana. It was oblivious. In the surge I accidentally brushed the creature with the dome of the camera. It kept on feeding, then raised its head as something inside its tiny dinosaur mind made a decision. It pushed off the rock, folded its little arms and legs back, and with its long tail sculled through the sea. Silhouetted against the surface, it looked like a living, swimming gargoyle that did not belong to our time or planet.

GALÁPAGOS SEA LIONS AT PLAY IN A SHALLOW SEA, SAN SALVADOR ISLAND, GALÁPAGOS

GALÁPAGOS SEA LION AND A SCHOOL
OF GRUNTS, DEVIL'S CROWN,
FLOREANA ISLAND, GALÁPAGOS

MARINE IGUANA, ABOUT TO BE TERRORIZED BY SEA LIONS, MUNCHING ALGAE, FERNANDINA ISLAND, GALÁPAGOS

MARINE IGUANA HEADING FOR HOME, FERNANDINA ISLAND, GALÁPAGOS

PALAU

Marine lakes are saltwater lakes surrounded by land but with a tenuous connection to the sea. Eil Malk, an island in the Palau chain, in the central western Pacific, enfolds several marine lakes in its green jungle arms. To reach the lakes it is necessary to climb through a thick, fecund jungle. With marine biologists William and Peggy Hamner, Anne and I climbed up the side of a ridge with our Palauan guides and porters and down the other side into a world where green replaced the blue of the sea. These lakes are warmer than blood, and hold life forms found nowhere else. Jellyfish Salt Lake is home to a vast school of small mastigias jellyfish that migrate up and down the length of the lake, following the path of the sun. We swam into the school, which slid around us, pulsating like yellow hearts.

In the Hot Water Lake the temperature was almost 105 degrees Fahrenheit. Anne and I swam along the edge of the lake under the branches of overhanging trees. The cold rain fell in fat drops that raised tendrils of steam as they hit the surface of the lake. We came around a corner and found a small saltwater crocodile lying on the surface, arms and legs outstretched, half-asleep. Its belly was a perfect unmarred light yellow. We looked at each other and for a moment shared a strange part of the Pacific. I fumbled at my camera, and the crocodile awoke and disappeared into the boiling green gloom of the lake.

Spooky Lake is another strange body of water in the center of Eil Malk. Here, bacteria form platelike layers in the water, the top layer clear, the next a cloudy green. The place looks like the moors of England where layers of fog cling to the ground. I watched Peggy Hamner dive into the primordial soup. The lake was silent and still. It needed the keening howl of a distant wolf.

PEGGY HAMNER IN FOGGY LAYERS OF WATER, SPOOKY LAKE, PALAU, MICRONESIA

MASTIGIAS JELLYFISH IN SUNLIGHT, JELLYFISH SALT LAKE, EIL MALK ISLAND, PALAU, MICRONESIA

KEREMA

GORGONIAN SEA FAN GARDEN,
KEREMA ISLANDS

RED CLOWNFISH IN BUBBLE-TIPPED ANEMONE, KEREMA ISLANDS

PANDA CLOWNFISH IN CARPET ANEMONE, KEREMA ISLANDS

SAND TIGER SHARKS, OGASARAWA ISLANDS

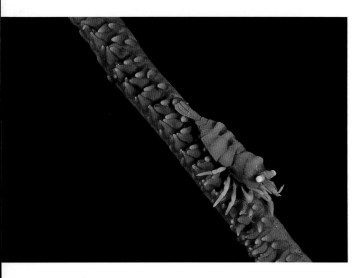

ONE-INCH-LONG ARMORED SHRIMP
ON WHIP CORAL, KEREMA ISLANDS

JAPAN

SURUGA BAY AND THE IZU PENINSULA

The snow-covered peak of Mount Fuji catches the dawn and glows a rich lacquer red that changes to rose with the growing light. The early light slips down the graceful shoulders of the mountain and touches the dark waters of Suruga Bay. The bay is a deep alcove of the abyssal ocean whose steep sides slide to a depth of nearly nine thousand feet at its entrance, a finger of the Pacific reaching out and touching the land of Japan. There is an almost mystical partnership between the sacred mountain and the huge, dark bay. The snow-covered peak holds the promises of gods; the dark depths are the realm of demons, home to such creatures as monstrous Pacific sleeper sharks and giant Japanese spider crabs, which can span twelve feet from claw to claw.

In the fall, winds from Siberia sweep over the slopes of Mount Fuji. On rare days the air over the bay clears, but usually Mount Fuji is hidden behind veils of mist and rain. The great mountain, lording over the bay, becomes a ghost.

In the fall of 1989 I joined my *Geographic* colleague photographer Emory Kristof, robot submersible inventor Chris Nicholson, and marine biologist Dr. Eugenie Clark's team, which included the famous Japanese underwater photographer Koji Nakamura, in Suruga Bay. It was a grand expedition that was to explore the bay from the very bottom to the sunlit shallows. Emory, Chris, and Genie took the depths and I got the shallows. Koji and his company, Japan Underwater Films, filmed the whole operation.

Our "shallow" group, which included Koji's partner Tadahiko Matsui, worked out of the little diving resort town of Osezaki. Diving in Japan is strictly limited to diving resorts or areas. In theory, all Japan's coastal seas belong to everyone. In reality, these seas belong to the fishermen of the hundreds of fishing villages that cluster around places like Suruga Bay and the cliffs of the rocky Izu Peninsula. The divers drive the fishermen crazy. The fishermen are convinced that the divers scare the fish, especially the lobsters. In truth the bay, like most of Japan's shallow coastal waters, has been wildly overfished. Suruga Bay, however, is so deep that it offers some protection to vast schools of pelagic fish.

On a day of mist, a late morning sun slid in and out of cloud layers as we anchored off the rocky, pine-tree-covered peninsula that separates the little harbor of Osezaki from the open bay. We were two teams with different missions. Koji, Matsui, and his assistants Okamura, Itoh, and Uryu would shoot film. My assistant, Nicki Konstantinou, and I would shoot stills.

We would make one long dive, going as deep as 170 feet, then slowly work up to the shallows, decompressing and shooting pictures. During the dive, which was to last over four hours, we would have to change tanks underwater. To keep warm we wore old-fashioned neoprene rubber wet suits with no zippers or nylon linings. I took ten cameras in housings, with lenses ranging from wide-angle to telemacro, and fifteen underwater electronic flashes. Koji and Matsui took the giant video camera and a half-dozen movie lights. We made a "dump," an assembly point at thirty-five feet, near the anchor rope, with our extra double tanks, cameras, and lights. Nicki and I selected the cameras we would need for the deep. Matsui prepared the video camera and the lights. We checked our air gauges, watches, and decompression meters. The water was green and clear with a slight layer of mist just under the surface from the night's rain. Matsui nodded and we took off, finning rapidly down the steep slope of Suruga Bay, flying into a dark land.

At 120 feet the wire coral forest began. Wire coral, a type of black coral, sprang twisting and curling from the bottom, reaching toward the light, feeding on plankton in the slight current. The forest grew thicker as we went deeper. Some coral strands were ten feet long. Schools of silvery cardinal fish flowed in and out of the forest. Cherry blossom anthias, found only in this secret corner of the Pacific, paraded in front of the cardinal fish like resplendent Kabuki actors, strutting in front of a silver curtain. The deep wire coral forest is a place of grand hallucination and illusion. Nicki's movie light shone like a headlamp on a midnight train. A lionfish hunted through straggly strands of coral that were silhouetted against the falling blue-green light from the distant surface. Lost in a Grimm brothers fairy tale, we wandered like children through a witch's scalp.

It was time to leave. The meters began to indicate a long decompression. We ascended, flying slowly out of the wire coral forest, leaving behind the strange dreams from this zone of nitrogen narcosis.

Mount Fuji last erupted in 1707, producing a thick blanket of volcanic ash and sand that is now full of life. Here I found a snake eel, a predator with armored eye slits and interlocking teeth. A long piece of a strange mucuslike substance clung to its nostril. Before I could examine it, the eel submerged itself in the sand. Another eel, yellow and about the size of my thumb, stuck its head out of the sand. It had white eyes and white spots and a toothless, muppetlike grin. I spent forty minutes creeping toward it with a long telephoto lens until it also disappeared in the soft sand, leaving tiny ripples. Dragonet fish hovered, swimming and levitating, grazing and gobbling the sand and expelling it through their gills. There is life between each grain, forming a gritty stone pasture full of microscopic crustaceans and worms.

At thirty feet Nicki found a perfect silver-dollar-sized juvenile lionfish. It had teal blue markings on its fins and swam like a butterfly in the open water. Matsui found an even smaller one that lay in his red diving glove and opened and closed its fins like a geisha's fan. We handled the lionfish carefully; even in this tiny state they are highly venomous.

The sand slope slowly gave way to a boulder-covered bottom. Koji and I found a pair of ghost pipefish swimming in tandem, the male hovering above the larger female.

69

These creatures are related to seahorses, but unlike the seahorse the female ghost pipefish broods her own eggs. They are perfectly camouflaged, clad in nubby, tempura-colored exoskeletons. For a moment I took my eyes away from the drifting, finning fish and they vanished against the complicated bottom.

In the third cold hour of the dive, I found a group of squid in ten feet of water, evil angels hanging just beneath the surface. Pale creatures composed more of water than flesh, they watched with their large silvery eyes for any movement on the stony bottom. An unsuspecting rockfish trundling from boulder to boulder was an easy victim. A squid descended, elongating its soft body, shot its tentacles forward, and enveloped the fish quickly and silently. Later, a shadow passed overhead. I looked up to the surface and came face to face with a "look-box" fisherman. Leaning over the side of his little skiff, he stared bleakly at me through a glass-paned look box, his window on the sea. I watched as he sent down a long spear that pinned another rockfish to the bottom and used a long probe with a tiny net to trap a crab.

Alone, I swam into the shallow water. The afternoon sun had broken through, tracing spidery light patterns on the bottom. I had been in the water for four hours, and the deep wire coral forest and the dark bay beyond seemed only an ancient memory.

The boulders in the shallow water were smooth and round, polished by the waves moving against the shoreline. They formed a seascape of utter simplicity, a temple garden, a carpet of algae-covered rocks rolled out of the sea through the curtain of the surface and up the slopes of land.

I climbed onto the dive boat, shivering and tired. Koji gave me a cup of thick, hot miso soup. In the growing afternoon shadows squid boats turned on their lights and rode on the surface of the great bay, waiting for nightfall beneath the shoulders of a sacred mountain. As the light fled the Pacific, Mount Fuji changed from an outline to a presence to a memory.

HOVERING SQUID, OSEZAKI, SURUGA BAY, JAPAN

A DEEP-DIVING ROBOT CAMERA —
NATIONAL GEOGRAPHIC'S ROV (REMOTE
OPERATED VEHICLE) *SEAROVER* —
BEGINNING ITS DIVE IN THE SHADOW
OF MOUNT FUJI, SURUGA BAY, JAPAN

HERMIT CRAB WITH HYDROID PLUME,
OSEZAKI, SURUGA BAY, JAPAN

LIONFISH IN WIRE CORAL FOREST,
OSEZAKI, SURUGA BAY, JAPAN

CARDINAL FISH SCHOOL HOVERING
OVER WIRE CORAL FOREST, OSEZAKI,
SURUGA BAY, JAPAN

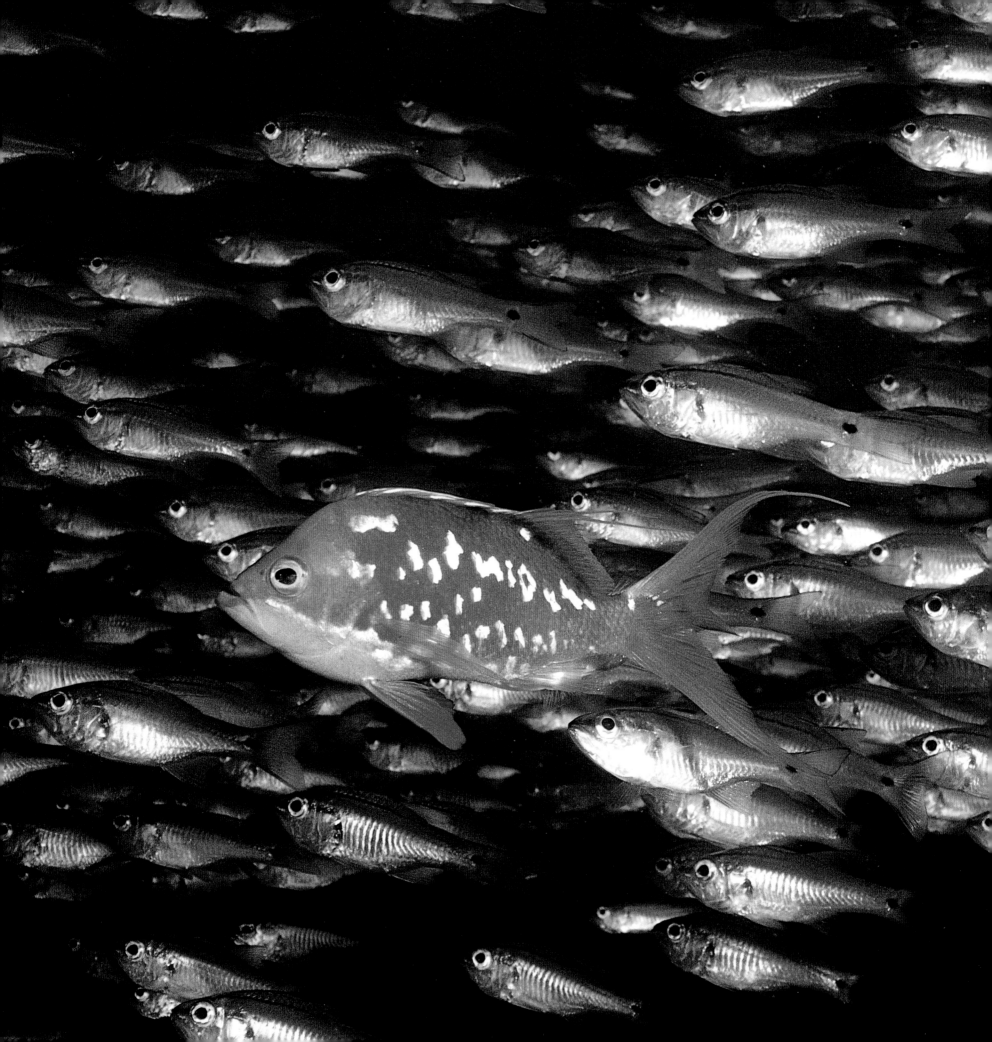

MALE CHERRY BLOSSOM ANTHIAS
SWIMMING IN FRONT OF A CURTAIN
OF SILVERY CARDINAL FISH, OSEZAKI,
SURUGA BAY, JAPAN

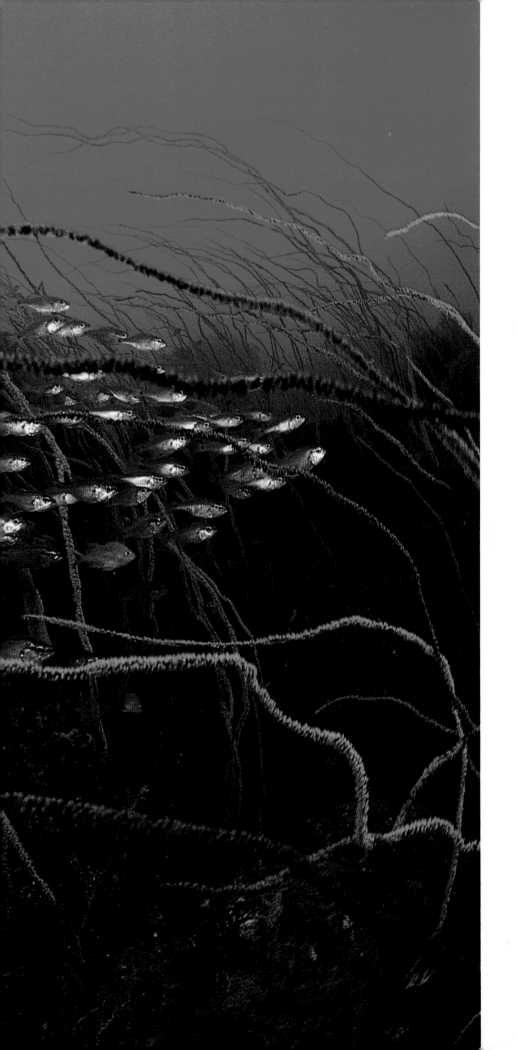

CARDINAL FISH FLYING THROUGH
WIRE CORAL FOREST, OSEZAKI,
SURUGA BAY, JAPAN

TIGER MORAY EEL, FUTO POINT, IZU PENINSULA, JAPAN

JAPANESE MORAY EEL SLITHERING THROUGH ORANGE SOFT CORAL, FUTO POINT, IZU PENINSULA, JAPAN   81

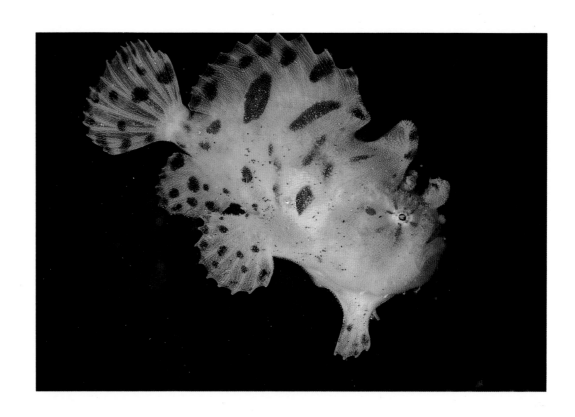

YELLOW FROGFISH, OSEZAKI, SURUGA BAY, JAPAN

CERIANTHID SAND ANEMONE,
OSEZAKI HARBOR, SURUGA BAY, JAPAN

SCORPIONFISH, OSEZAKI HARBOR, SURUGA BAY, JAPAN

MALE (LEFT) AND FEMALE (RIGHT) GHOST PIPEFISH, OSEZAKI, SURUGA BAY, JAPAN

A JAPANESE MORAY EEL IN ITS HOME
IN A TRUCK TIRE, OSEZAKI HARBOR,
SURUGA BAY, JAPAN

YELLOW "MUPPET" SNAKE EEL, OSEZAKI,
SURUGA BAY, JAPAN

SNAKE EEL WITH MUCUS, OSEZAKI,
SURUGA BAY, JAPAN

ROCKS AND SURFACE LIGHT, OSEZAKI,
SURUGA BAY, JAPAN

88

OVERLEAF:
EVENING, MOUNT FUJI, AND SQUID BOATS,
SURUGA BAY, JAPAN

JAPANESE FISHERMAN USING A LOOK-BOX,
OSEZAKI, SURUGA BAY, JAPAN

THE PACIFIC WAR

From forty-one thousand feet the Pacific is shrouded by whirlpools of nimbus clouds, soft gray galaxies of moisture that emerge in the growing light. I was flying westward to Australia, cocooned in a Qantas 747, pursued by the dawn. The cabin was dark and the other passengers rustled to squeeze more sleep out of a very short Pacific night. I looked out my window with mild, heavy eyes, drifting at the edge of consciousness. I dreamed.

Forty-five years earlier and thirty-thousand feet below another aircraft held the same course over the South Pacific—a Boeing B-17 Flying Fortress, new from the factory. The olive-drab paint was beginning to flake off in the driving rain of the swirling Pacific storm. Inside was controlled chaos. The four huge radial engines roared against the storm while the entire aircraft vibrated, "a hundred thousand rivets flying in formation." The pilot, an old man at twenty-five, fought the controls as the aircraft rose and fell a hundred feet at a time. Drops of sweat raced down the length of his arm, across his wrist and palm onto the black crinkle finish of the Boeing control wheel. Though a B-17 appears impregnable, the Flying Fortress is really a series of aluminum ribs covered with a thin aluminum skin. In places you can stick a screwdriver through. It's like being in the belly of a cheap metal fish. The plane smelled of aviation gas, the fumes seeping out of a huge tank that for the Pacific journey replaced the bombs in the bomb bay. The B-17 must have been full of fear, drowned out by the engines and the storm—the fear of ten "men," teenagers mostly, going off to war, lost in the darkness of a great Pacific storm. The B-17 was destined for New Guinea, where it would be part of the Pacific War.

DECEMBER MORNING, PEARL HARBOR, 1990, WITH SNJ TRAINING PLANE —
WHERE THE PACIFIC WAR BEGAN

The 747 made a midcourse correction. I awoke. The dream was a prediction — within a month I would dive on a downed B-17 that had crashed into the sea near Cape Vogel, on the southeast corner of New Guinea. Throughout the war, thousands of aircraft would cross the belly of the Pacific, island hopping, with simple navigational leaps of faith. This plane survived almost intact; resting on a seabed 150 feet deep, it looked as if it were flying through an endless dark blue dawn.

The instant a plane or a ship crosses the barrier between air and water, a strange thing happens. The pain, the fire, the smell, and the fear vanish. The sea enfolds, embracing the victim of war. It isolates and removes it from its human context, turning it into a sculpture that stands and stretches in the clear gallery of the Pacific. Then the sea does something more. It coats the sculpture with life, as an oyster coats an irritant with pearlescent nacre to produce a pearl. It is a softening process. Guns, shells, and the flanks of ships are covered with sponges. Cardinal fish swim through bomb bays, and gun sights become soft and fuzzy with red encrusting sponges. The sea begins the process of digestion.

The Pacific War was a war fought at arm's length. The power and the might of America and Japan had to stretch across the largest ocean on the planet, yet the battlegrounds were strangely small.

I took the fifteen-minute flight from Henderson Field on Guadalcanal to the islands of Florida and Tulagi in a Hughes 500 helicopter, across the southern end of the Slot, a passage formed by the Solomon Islands. In the fall of 1942 this little place in the Pacific was filled with Japanese and American aviators fighting for their lives. Japanese bombers and fighters from the great naval base at Rabaul on the eastern tip of New Britain Island, five hundred miles to the northwest, would fly down the Slot. Australian coastwatchers hiding in the islands' mountainous jungles would radio "one hundred planes headed your way" to the Americans at Henderson Field, who would launch their fighter planes into the air, clawing for the advantage of altitude. The American marines flew Grumman F4F Wildcats, simple, almost generic fighter planes so basic that the landing gear had to be cranked up by hand. They were no match for the Japanese Mitsubishi Zero, an aircraft that could fly five hundred miles, fight, and return. The Zero's great secret was its remarkable magnesium-capped wing spar, which made the plane light, strong, and as flexible as a Japanese sword, able to outturn, outclimb, and outshoot the stubby Wildcat. The Zero was the best plane in the world, but it was an offensive weapon; it had no armor.

I looked out the open doorway of the helicopter at the little amphitheater of islands, the hulk of Guadalcanal, the maze of Florida and Tulagi, and to the northeast the lump of Savo Island. It was a stage set rimmed with brilliant clouds. The Japanese came, Zeros and fat, cigar-shaped Betty bombers (another Mitsubishi product), to destroy the Americans at Henderson Field. The Wildcats would fall down on their formation, shoot at the Japanese bombers and fighters, then keep diving toward the sea, never daring to dogfight with the deadly, agile Zeros. It all happened in a place smaller than New York Harbor, a place of glowing green islands and a rich blue sea.

The afternoon sun went behind a small rain squall, creating shafts of light that bathed dark Savo Island in a yellow glow. In the fall of 1942 the Americans' Wildcats barely held on to the day. The night belonged to the Japanese. Expert at night maneuvers, their destroyers, cruisers, and battleships followed their air fleets down the Slot. In the Battle of Savo Island, August 1942, the Imperial Japanese Navy sank one Australian and three American cruisers in a single night, the worst naval defeat in American history. Their secret was a remarkable weapon, the Long Lance torpedo, launched from five miles away. The only indication of the torpedos was the bioluminescent wake they made as they stretched for their targets. The ships went down behind Savo Island in very deep water. Despite strong currents, the surface here can be mirror-smooth, and oil from the sunken ships seeps upward. This place in the sea, this grave of thousands, is called Ironbottom Sound.

That night I went northwest from Guadalcanal to the Russell Islands. There was a sliver of moon, wisps of clouds, and a flat sea. I stood on the bow of the dive boat. Savo Island and the coastline of Guadalcanal were only heavy shadows in the night sea. Suddenly a phosphorescent green wake streaked toward the boat. Not an imaginary nightmare torpedo—it was a dolphin. The animal curved around and caught and rode the bow wave of the boat. Its whistling breath sounded like a sigh.

The Pacific War produced instant cities on the edge of a primitive ocean wilderness. At times the Japanese had a quarter of a million troops in the fortress of Rabaul. On Espíritu Santo in the New Hebrides (now Vanuatu), the Americans built a huge base, a staging place for the Pacific War. There were at times ten major hospitals, thirty-three movie theaters, and a quarter of a million men. Now the main base is a small town with wide streets and six Chinese restaurants. In the heat of the day, dogs sleep in the middle of a dusty boulevard.

The sea at the edge of town is the resting place of the grandest shipwreck in the South Pacific—the *President Coolidge,* a luxury liner turned troop transport. On October 26, 1943, the *Coolidge* came up the wrong channel and struck an American mine. Somebody didn't "get the word." The mine ripped a hole in the engine room and the captain drove the twenty-two-thousand-ton ship onto a reef. An entire division—5,440 men—got off the ship; then the liner ponderously turned on its side and slid off the reef. The Forty-third Division lost everything—guns, uniforms, helmets, ammunition, and food—yet, amazingly, only three people died in the shipwreck.

I dove on the *Coolidge* with Allan Power, an Australian diver and underwater photographer who came in 1971 with a salvage group, fell in love with the great ship, and stayed as self-appointed curator of this silent blue museum. Allan showed me the ship. Photographers Kevin and Chris Deacon from Sydney dove and helped with the lights. We were like children exploring an ancient, empty iron castle. The monstrous holds were full of jeeps and guns, stirred into a metal bouillabaisse. The promenade deck was littered with helmets, gas masks, and Garand rifles. We swam through the darkened engine room, then past the ship's boilers. In one hold, Allan showed us drop tanks for P-38 fighters that looked like alien eggs. The sea had covered a Thompson submachine gun with marine growth, turning it into a soft toy. We found the Forty-third's typewriter, also covered with sponges—more oceanic revenge. In the very bowels of the ship, over 130 feet down, Allan led us into the strangest room in the South Pacific—the portside enlisted men's head. Row after row of toilets stretched into the gloom. We swam up over the flanks of the ship, past the bridge and the port gun tub, full of silversides, up toward the bow, then to the reef—the shallow reef full of sunlight and fish. The *Coolidge* disappeared, a memory beneath our fins.

In the Russell Islands north of Guadalcanal, I made an amazing historical discovery. For every shell fired, someone drank a Coke. I saw mountains, reefs, shoals of Coke bottles off Mbanika Island to match pallet loads of antiaircraft shells that covered the lagoon floor as far as I could see. At the end of the war the Americans loaded pallets of shells onto barges and bulldozed them into the sea. When the job was done, they drove the bulldozers themselves off the barges. We found one bulldozer sitting upright 160 feet down. Kevin climbed into the driver's seat and tried to start it. It was flooded.

In Simpson Harbor, Rabaul children played in the shallow water around the wreck of a Kawanishi flying boat. Fish swam about the cockpit of a Japanese Zero. The 560-foot-long *Hakai Maru,* a naval repair ship sunk in 1943 by American naval aircraft, sat upright on a dark volcanic sand bottom, steaming forever through a blue undersea fog on a journey to nowhere. Rabaul is surrounded by five active volcanos. The town is on perpetual alert. Tendrils of white smoke leak from a volcano overlooking the airport. During the war, Japanese pilots said the place smelled like hell from the sulfurous volcanic smoke and round-the-clock bombing by American and Australian forces. Rabaul was cut off and bombed but never invaded; the Japanese had dug three hundred miles of tunnels beneath the city.

The sunken planes and ships of the war exist in a timeless netherworld of the sea, but strangely they are the most powerful reminders of the Pacific War. The cities by the sea—of Guadalcanal, Espíritu Santo, Milne Bay, Manus Island—were born and died in only a few years. Millions of people went through these cities in the process of war, yet none looked beneath the surface. The war raged on over the richest coral environment on our planet.

At 7:55 A.M. on December 7, 1990, I stood on board the *Arizona* Memorial at Pearl Harbor in the Hawaiian Islands. Every December 7, there is a service held at the memorial honoring the war dead, remembering Pearl Harbor. A kind of bridge, the memorial straddles the sunken battleship *Arizona*. Forty-nine years ago, a 1,700-pound naval shell that had been converted to an aerial bomb was dropped on the *Arizona* from a Kate high-level bomber. The bomb penetrated the forward magazine of the *Arizona;* in one cataclysmic explosion, 1,177 sailors died instantly.

The day before the memorial service, I dove on the *Arizona* with Dan Linehan of the National Park Service. He runs the Submerged Cultural Resource Unit—the curators of the *Arizona* and other important historical ships. We swam along the green, gloomy sides of the ship. The visibility in Pearl Harbor is about six feet on a good day. The sea had taken over the battleship. Nothing was recognizable. We then came to the forward guns. The guns from the Number One turret had never been salvaged. I swam under them and looked up. They were immense, dark, deadly fingers silhouetted against the green surface light. The Pacific War, World War II, for Americans had rolled for three and a half years across this vast ocean, leaving death in its wake. Fifty years later light still streams into the sea, a sea which remembers nothing, yet for a brief time preserves everything.

It's 8:00 A.M. The colors have been raised on the *Arizona* Memorial. A destroyer in the harbor fires a salute. I look westward down Pearl Harbor toward Hickham Air Base and Honolulu International Airport's reef runway. In the sky at the edge of Pearl Harbor five shining silver 747s are on final approach. They have come eastward, riding the jet

GUNS OF THE *ARIZONA*, PEARL HARBOR, HAWAII

stream out of a Pacific night. Inside the great aluminum tubes flight attendants request in Japanese, in soft bell-like tones, that tray tables and seat backs be returned to their upright positions. I look over the side at the dark shadow of the *Arizona*. A drop of bunker oil, entombed for forty-nine years, rises upward. When it touches the underside of the surface, it spreads outward, reflecting the morning light like a rainbow.

FIVE-INCH GUN AND THREE-INCH GUNS ON THE WRECK OF AN AMERICAN FLEET OILER,
GUADALCANAL, SOLOMON ISLANDS

IRONBOTTOM SOUND WITH RUSSELL ISLANDS AND GUADALCANAL IN BACKGROUND, SOLOMON ISLANDS   103

WRECK OF KAWANISHI FLYING BOAT,
SIMPSON HARBOR, RABAUL,
PAPUA NEW GUINEA

MITSUBISHI ZERO, SIMPSON HARBOR,
RABAUL, PAPUA NEW GUINEA

BOEING B-17E FLYING FORTRESS, CAPE VOGEL, BOGA BOGA, PAPUA NEW GUINEA

LOCKHEED P-38 WRECK, MILNE BAY, PAPUA NEW GUINEA

FORWARD HOLD, *PRESIDENT COOLIDGE*,
ESPÍRITU SANTO ISLAND, VANUATU

ENLISTED MEN'S HEAD, *PRESIDENT
COOLIDGE*, ESPÍRITU SANTO ISLAND,
VANUATU

STARBOARD GUNTUB, WITH
SILVERSIDES, *PRESIDENT COOLIDGE*,
ESPÍRITU SANTO ISLAND, VANUATU 111

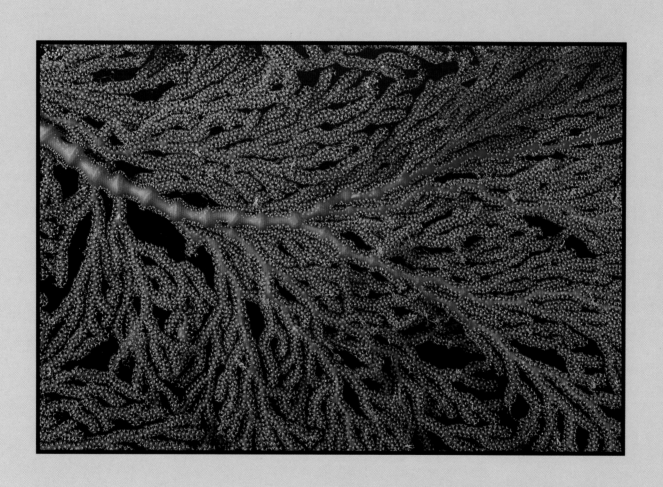

PAPUA NEW GUINEA

Bob and Dinah Halstead backed their dive boat *Telita* into the jungle. The deep water of the Goschen Strait at the edge of the Solomon Sea runs right up to land. In the wheelhouse the fathometer frantically tried to gauge depths beyond its range until suddenly, as *Telita* swung over the top of the submarine cliff, it registered ten feet. The crew tied *Telita* to a tree and Bob put down a bow anchor. We were moored to the side of the enormous island of New Guinea, near its eastern tip on the outer arm of Milne Bay.

This is the gateway to a coral universe, the warmest water, with the greatest number of species of marine life, and the richest coral growth on our planet. Westward, the Pacific divides up into complicated, convoluted ocean corridors, passageways, and seas with mysterious, exotic names: the Banda Sea, the Java Sea, the Celebes Sea, and, at the farthest reaches of the Pacific, the South China Sea. The island of New Guinea is a land unto itself, with mountain ranges, rain forests, and long, silt-laden rivers that wash nutriments into the coral-rich seas. Usually, close to land, these waters are muddy and visibility can be measured in inches. But in the right season the ocean lies flat and water and land touch with the delicacy of an Adirondack lake. Clean currents from the Solomon Sea brush against the land.

I stepped off the stern of *Telita* in clear, deep water. A field of chalice corals formed a series of blue-green steps along the edge of the cliff, where colonies of purple-tipped anemones grew. I swam along a deep channel gouged out underneath the bank of land. Trees from the rain forest grew out over the bank to shade the sea, creating a shadowy

GORGONIAN SEA FAN, MILNE BAY, PAPUA NEW GUINEA

twilight that allowed a garden of gorgonian sea fans to flourish. Tiny gobies live out their lives among the branches of the sea fans. A tree from the rain forest touched the water's surface and a baby batfish swam among its branches. In a nearby cove on a volcanic, sandy slope, I found an orange-and-yellow fire urchin bristling with venomous spines. A school of small black fish orbited over the top of the urchin and pairs of two different species of shrimp wandered among its spines and wiggling tube feet. The coral Eden of the southwest Pacific harbors castles of life, larger forms that support and nurture the smaller, more delicate creatures.

Anne and I and my sister Jane swam among the paths of this reef at the edge of New Guinea. We were in a distant part of the Pacific, yet at home in the sea. This was Dinah Halstead's land. She grew up here. Our daughter, Emily, had her third birthday here. It is a garden, not a jungle.

At the edge of Milne Bay near Basilaki Island the sea is strewn with hundreds of small reefs and rocks. Few have been explored. Bob edged *Telita* up to a likely virginal-looking reef and put the anchor down on the upcurrent face. We went into the water. The current was strong, pushing against the face of the reef then sliding over the top like air across an aircraft's wing. On the top, though the reef was scoured by currents, schools of purple anthias hovered above it. At the edge the current passed over us and the reef sloped gently downward, covered by row after row of giant yellow sea fans. Millions of silvery baitfish formed rolling clouds above the sea fans, boiling and churning as hunting jacks and tunas speared into the tremendous schools. I watched Bob and Anne soar beneath the clouds and over an endless waving field of sea fans. Parades of unicorn fish passed below, feeding in the open water on the face of the reef. Deeper still, gray reef sharks patrolled. At the end of the dive I swam up over the face of the reef and flew with the current back to the boat. As usual, I was the last out. Anne said, "Bob wants to see us in the wheelhouse."

Bob glanced at the numbers on his satellite navigation system, then unrolled the old

SEA FANS, DOUBILET'S REEF, MILNE BAY, PAPUA NEW GUINEA

British admiralty hydrographic survey chart of the Milne Bay area. He took a perfectly sharpened yellow pencil (a navigator's secret delight) and with tiny, precise calligraphy marked down the satellite numbers where the chart indicated a rock. Then, carefully, he wrote, "Doubilet's Reef."

DAMSELFISH AND GLASSY SWEEPERS, "REEF AT THE EDGE OF THE WORLD,"
MILNE BAY, PAPUA NEW GUINEA

RAIN FOREST AND SEA, MILNE BAY, PAPUA NEW GUINEA

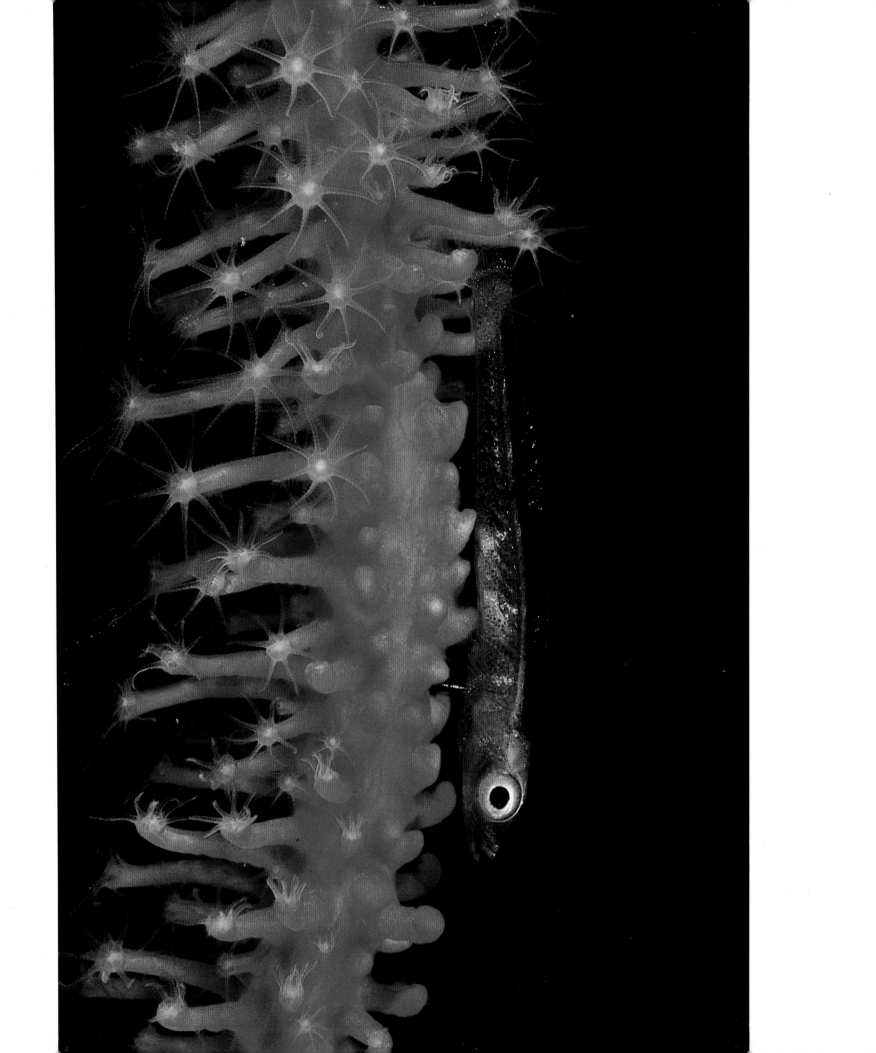

ORANGE CLOWNFISH IN ANEMONE, DOUBILET'S REEF, MILNE BAY, PAPUA NEW GUINEA

GOBY ON WHIP CORAL, DINAH'S LAND, MILNE BAY, PAPUA NEW GUINEA

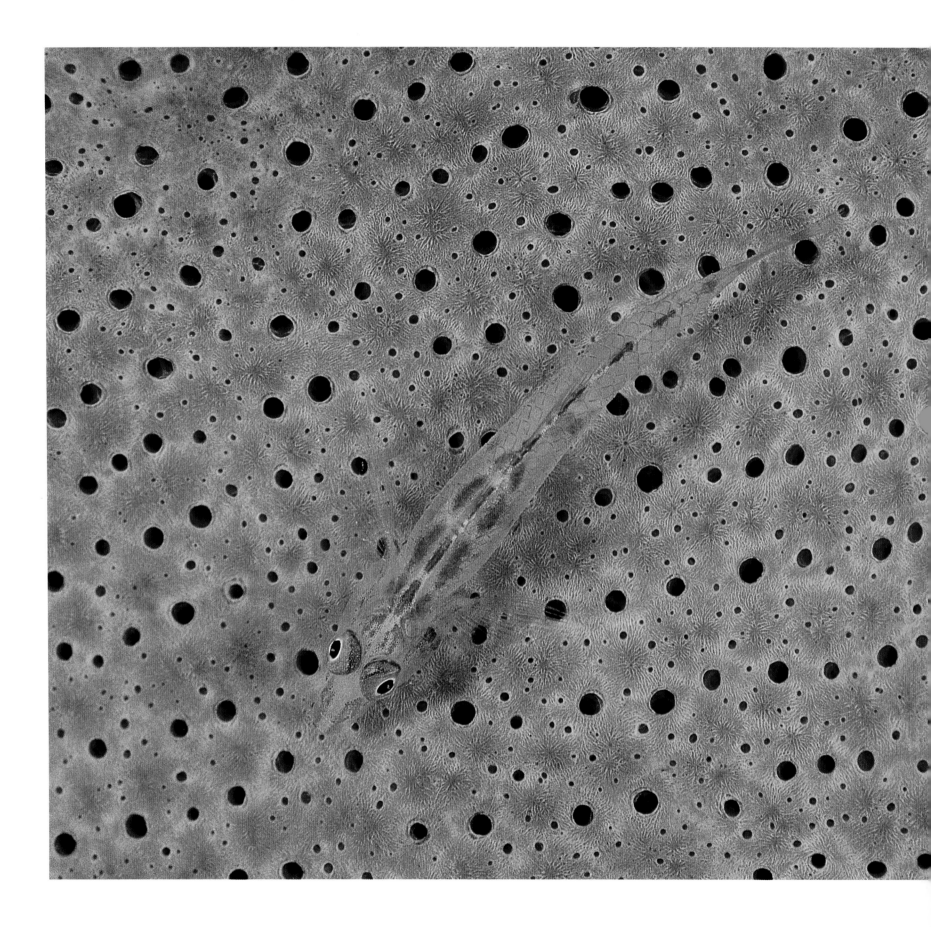

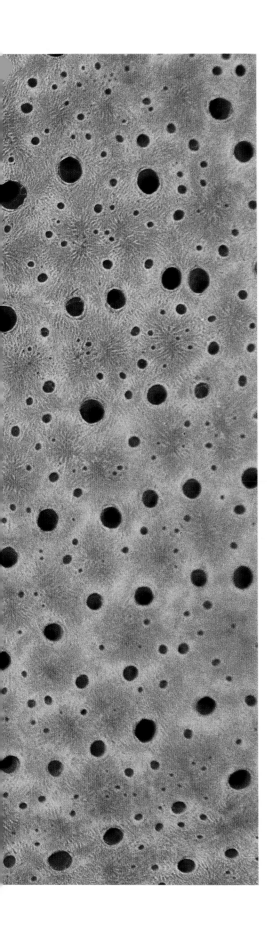

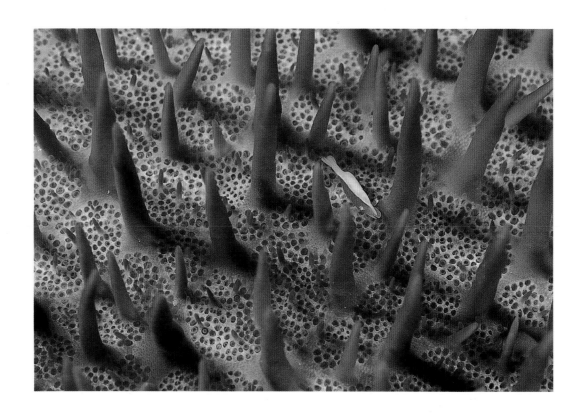

SHRIMP WALKING BETWEEN SPINES
OF A CROWN-OF-THORNS STARFISH,
RABAUL, PAPUA NEW GUINEA

GOBY ON ELEPHANT-EAR SPONGE,
MILNE BAY, PAPUA NEW GUINEA

PACIFIC BARRACUDAS, NEW HANOVER ISLAND, PAPUA NEW GUINEA

DANCE OF THE BARRACUDAS, NEW HANOVER ISLAND, PAPUA NEW GUINEA

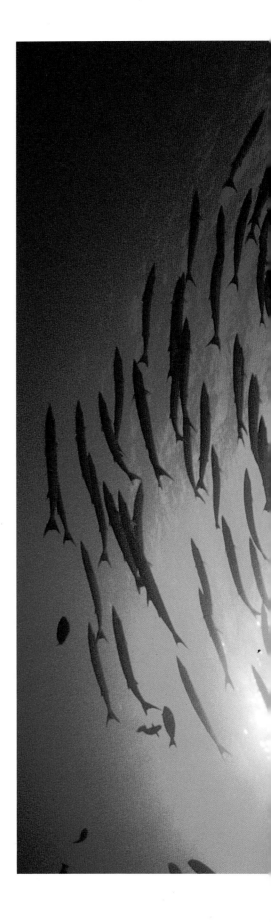

YELLOW-FACED ANGELFISH, NEW IRELAND ISLAND, PAPUA NEW GUINEA

UNICORN FISH, NEW HANOVER ISLAND, PAPUA NEW GUINEA

PURPLE ANTHIAS AND YELLOW
DAMSELFISH SWIM PAST CHALICE CORAL,
MILNE BAY, PAPUA NEW GUINEA

126

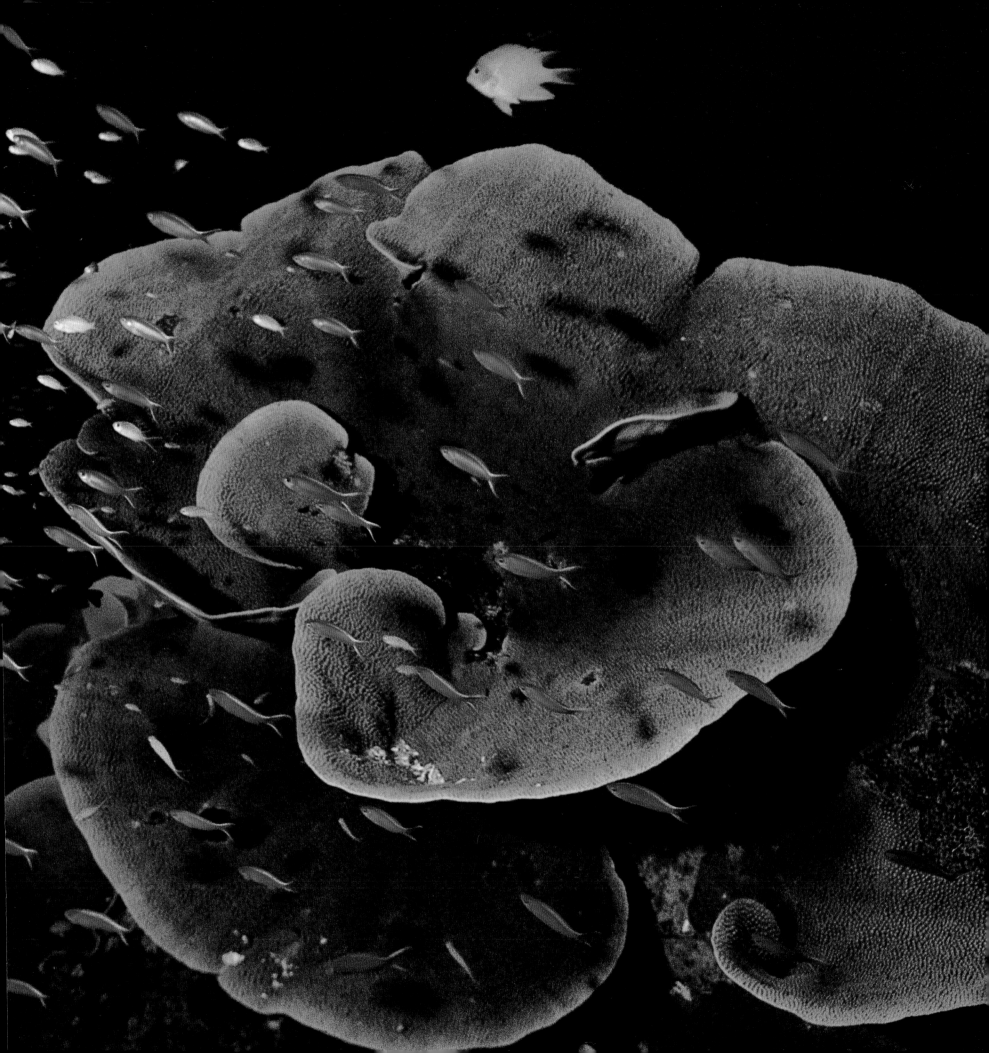

THE GREAT BARRIER REEF

As we ran south from Raine Island to the Great Detached Reef, the seas became mountainous. The southeast wind, blowing from the Coral Sea and the open Pacific, raised huge swells that passed between the breaks in the reef. Stubby little *Auriga Bay* rolled through the swells with vicious snapping movements.

Anne's face turned the color of glistening green-apple jade. She took to the bunk, leaving instructions for me to take Emily on deck to get some fresh air. With my feet pushed against the rail and my back against the wheelhouse, I clutched life-jacketed Emily in my lap. We looked bleakly at the large, rolling, blue beam seas as they coursed alongside *Auriga Bay*. One slip and Emily would be over the side in a second. Squeezing her, I turned to Graham McCallum, *Auriga*'s skipper. I said, "I'm scared all of the time here." He answered, "What the hell's the matter with you? *You* live in *New York City!*"

Another wild place. But the Great Barrier Reef of Australia is an oceanic wilderness, sometimes hundreds of miles from land. A coral country that stretches over 1,250 miles along the north coast of Queensland, the reef is not a single wall but a series of unconnected reefs, coral pinnacles, and endless coral mazes. The Number Ten Ribbon Reef is the classic of the outer reefs, a bastion facing the open ocean, stretching for miles like a great snake uncoiling in a dark sea. The Great Barrier Reef derives its power and its wealth not from the open ocean but from the coastal rain forests of North Queensland and the Cape York Peninsula. It is unique because it has what marine scientists call a lagoon: a body of stable, thick, nutriment-rich green water trapped between the outer reefs and the coast. Large

CLEANING WRASSE PICKS AT PARASITES ON A CORAL TROUT, COD HOLE,
NUMBER TEN RIBBON REEF, GREAT BARRIER REEF, AUSTRALIA

tides wash water over the reef while feeding open sea water into the lagoon. The Barrier Reef blooms, but the water is not crystal-clear, as it is in the Coral Sea. Isolated Marion Reef, two hundred miles out to sea from the outermost part of the Barrier Reef, has ultraclear water, big pelagic fish, and many sharks, but the coral growth and species diversity is sparse compared to the Barrier Reef.

In December 1987, Anne and I brought Emily on a long, rough northern journey down the Barrier Reef from Raine Island to the town of Cairns. The unpredictable weather makes diving the Barrier Reef, especially the distant northern end, a gamble. We sort of lost, but in the process saw some extraordinary things.

At Raine Island on certain spring nights, thousands of green turtles come ashore to lay eggs. We missed it, but we found some turtles sitting in a kind of coral parking garage all along the reef front, waiting until nightfall to crawl ashore. At Number Eight Sand Cay, 150 miles south, Emily and I found a big green turtle in the process of laying her eggs. We scooped sand away from her back flippers and we watched the heavy, mucus-coated, Ping Pong ball–sized eggs fall softly into the warm sand nest. The next day, we found hatchlings from another nest emerging, then racing for the sea. Emily took up one of the hatchlings. It was the same size as her hand.

On the northern end of Number Ten Ribbon Reef lies the fabled Cod Hole, a little coral amphitheater at the inside edge of the reef. The actors are a troup of ten or more 150-pound potato cod, wonderfully expressive, aggressive creatures that all look like Charles Laughton. Famed Australian underwater photographers and naturalists Ron and Valery Taylor discovered Cod Hole in the mid-1970s and spent a decade trying to preserve it. The friendly doglike fish would have been easy victims for any fisherman.

Graham McCallum went down with a plastic bin full of fish. The cod mobbed him. A definite pecking order was established: the larger fish hovered millimeters from Graham's ears, while the smaller hung underneath his arms. Graham removed a piece of fish from the plastic container—pandemonium, a weightless waltz of gluttony. The cod inhaled their food. The mouths would open, creating a whirlpoollike vacuum, and in would go five to ten pounds of fish. For an hour we fed them. Their stomachs grew hard, rounded and distended. The small ones seemed insatiable. The poor neglected creatures, long relegated to the back of the bread line, were in heaven.

When the cod were full they meandered along the reef like lost barrage balloons. There were a few half-hearted attempts at sex play. One cod would bump another and they would both blush white. But love was not to be. Indigestion came first, then cleaning. Teams of blue-and-white striped, three-inch-long cleaning wrasses set to work on each bloated fish. The wrasses would go in through their gills while the cod yawned, exposing rows of tiny teeth and gill rakers. The wrasses would pick scraps of food from between the teeth. The big fish would blush, change color, and sometimes lie on the bottom in complete surrender while the cleaning wrasses gave them a thorough going-over. Then one by one the cod left the amphitheater to return to their coral lairs and await the next visitors bearing gifts from above.

Heron Island, seven hundred miles south of Cod Hole, is one of the most southerly parts of the Great Barrier Reef. A resort island owned by the P&O Steamship Company, it usually has about three hundred paying guests, joined by one hundred thousand white-capped noddy terns plus forty thousand wedge-tailed shearwaters, and of course, the odd heron. The island is part of the Capricorn Group, a miniarchipelago completely different from any other part of the reef. Composed of tree-covered sandy islands and surrounded by large palette-shaped coral reefs, from the air it looks like aquamarine thumbprints on a dark ocean.

The Bommie, a large boulder-shaped coral head at the entrance to the Heron Island boat channel, is another famous place on the reef where humans interact with fish. Islanders have been feeding fish at the Bommie for decades. The Bommie compacts life on the reef into a smaller place, making it easier to observe courtship, territorial assertiveness, feeding, and cleaning—the secret lives of fish. I have always felt a little guilty about fish-feeding areas in the sea. There is something inelegant about it. Feeding disturbs the fishes' dignity, but at least *we* are not eating *them*.

I waited and watched for hours in shallow water next to the Bommie. I saw harlequin tuskfish fighting, venus tuskfish building nests and digging for food, batfish schooling and feeding on detritus, schools of silvery jacks sweeping along the bottom, and sweetlips opening their grand mouths to be cleaned by wrasses. It was a condensed version of the entire universe of the Barrier Reef.

At night, the terns settled into the trees of Heron Island. We put Emily to sleep and turned in. Suddenly Anne sat bolt upright and screamed, "What's that noise! That moaning? Is it Emily?"

Shearwaters, sometimes called mutton birds, had dug burrows in the sand and proceeded to court each other with low, mournful moans. The night world outside our Heron Island cottage was full of cheap 1950s horror-film moans, groans, and phony wolf howls. For the shearwaters it was music and poetry, the ancient expression of lost love and desire. The birds had migrated the length of the Pacific Ocean to reproduce on the tiny dot of Heron Island, at the very end of the Great Barrier Reef.

SAND CAY AND SEA BIRDS, NORTHERN GREAT BARRIER REEF, AUSTRALIA

GREEN TURTLE HATCHLING,
HERON ISLAND, GREAT BARRIER REEF,
AUSTRALIA

JACKS HUNTING BAITFISH, GREAT DETACHED REEF, GREAT BARRIER REEF, AUSTRALIA

SWEETLIPS, RAINE ISLAND, GREAT BARRIER REEF, AUSTRALIA

COD HOLE, NUMBER TEN RIBBON REEF, GREAT BARRIER REEF, AUSTRALIA

CLEANING WRASSE AND POTATO COD, COD HOLE, NUMBER TEN RIBBON REEF, GREAT BARRIER REEF, AUSTRALIA

EMPEROR FISH, GREAT BARRIER REEF, AUSTRALIA

EMPEROR FISH AND SNAPPERS,
THE BOMMIE, HERON ISLAND,
GREAT BARRIER REEF, AUSTRALIA

REEF AND BREAKING WAVES, LIZZARD ISLAND,
GREAT BARRIER REEF, AUSTRALIA

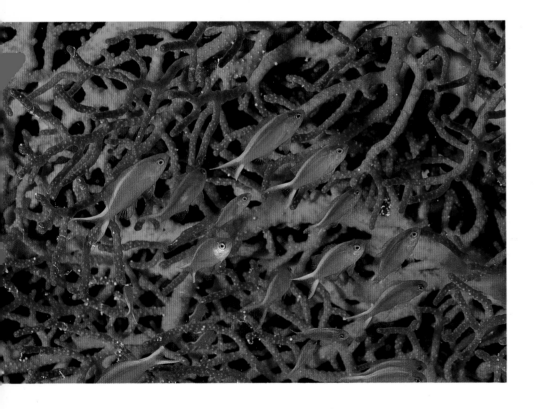

FAIRY BASSLET AND SEA FAN,
GREAT DETACHED REEF, GREAT BARRIER REEF,
AUSTRALIA

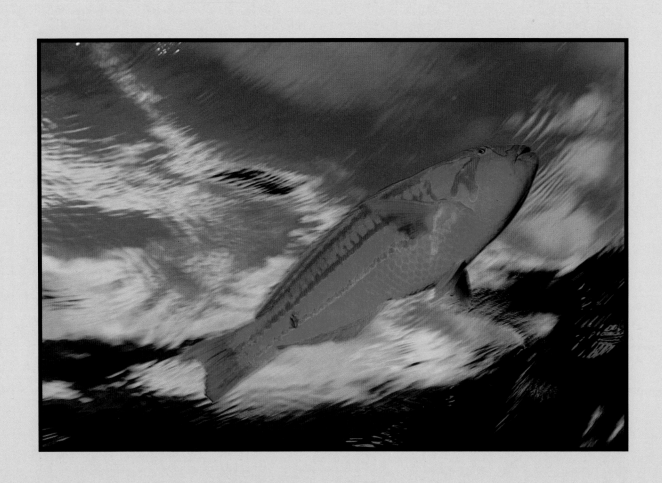

LORD HOWE ISLAND

Emily peered out of the little twin-engined Beechcraft's window. We descended out of an infinite, blue, cloud-veined Pacific sky, and banked over Lord Howe Island. Emily turned to us and said, "It looks like never-never land, Peter Pan's island." Lord Howe is a dream of an island, a perfect place with two 2,500-foot mountains that catch the Pacific wind and cradle a delicate cool rain forest. A temperate island of northern summer surrounded by a clear sea, it lies at the very edge of the chilly Tasman Sea but is bathed in a warm equatorial current. The coral reefs that fringe it are the most southerly in the world.

At 370 miles off Australia's eastern coast, Lord Howe is alone in an empty ocean, not part of an archipelago but a sliver of land left by a mile-high volcano that erupted seven million years ago. It was discovered (truly discovered, because it was uninhabited) in 1788 by Henry Lidgbird Ball, commander of the British Navy Tender HMS *Supply*. It was a peaceable kingdom where the creatures showed no fear of man. They were eaten—as an antidote to ghastly British naval cuisine. The island was settled in 1834 and the thin, strong, native Kentia palm trees were later exported for use in the palm courts that graced the interiors of great American and European hotels and legendary steamships. Tourism started in the early 1900s. After World War II, Short Sandringham flying boats began to service the island. By the late 1970s an airstrip was put in and the flying boats were retired. In 1982 Lord Howe Island was placed on the World Heritage list and development was halted. There are only about 280 residents, with beds for 390 tourists, on the seven-mile-long island. Lord Howe Island is an oceanic island, an oasis in an empty sea, uninfluenced by the weight of the Australian continent or the distant Great Barrier Reef.

The little Beechcraft turned on final, lowered its flaps, and sailed onto the airstrip that bisects the island. The pilot shut down the engines and they ticked and creaked as they cooled. A cow grazing near the airstrip lowed and the sea rumbled against the distant reef.

Island dive guide Jeff Deacon and my assistant, Gary Bell, met the plane. We piled the gear into the back of Jeff's van and drove along the lagoon road to our guest house. Jeff looked out to the sea past the lagoon. "It's your lucky day," he said. "The wind's been blowing for two weeks and it just stopped this morning. We're in the middle of the ocean, don't you know."

We dove with Jeff in the clear sea beneath the sheer face of Mount Malabar. Red-tailed tropic birds soared along the edge of the cliffs. The reef of big plate corals stair-steps down to a deep, sloping white-sand shelf. It is a simple reef, not overgrown and full of visual confusion like the Great Barrier Reef or the reefs of New Guinea. The endemic fish here are conservatively dressed. The Lord Howe Island clownfish are black, brown, and white, not the orange or yellow of their tropical relatives. I found a foot-long spotted hawkfish, resting on its pectoral fins. A pair of bespectacled angelfish passed rapidly by. Squiggly lines around the eyes make the creature resemble a surprised Prussian count with a monocle. The shadow of Mount Malabar crept darkly over the seabed. Jeff found a large Spanish-dancer nudibranch feeding on a sponge. It lifted off its meal, undulating, dancing in the soundless ballroom of an evening sea.

At Erscott's Hole in the lagoon I watched a Lord Howe Island double header wrasse pulverize a sea urchin. It flipped over the urchin and repeatedly smashed its tusklike teeth into it. A cloud of smaller pink, orange, and yellow wrasses orbited around the notoriously messy eater, waiting for scraps of urchin guts. The shallows of the lagoon were full of the life that huddles around the flanks of Lord Howe Island, beneath the peaks of the twin mountains.

DOUBLEHEADER WRASSE EATING A SEA URCHIN, ERSCOTT'S HOLE, LAGOON, LORD HOWE ISLAND, AUSTRALIA

North of Lord Howe Island two reefs, Elizabeth and Middleton, grow out of the seabed, part of the Lord Howe Rise, a kind of blue highway that wanders north all the way to New Caledonia. I chartered a twin-engined Piper aircraft, took the door off, and flew for an hour and a half up to distant Middleton Reef. It was an absolutely lonely flight. I stared out the open doorway into an empty, endless Pacific as the plane seemed to paddle ineffectively through an ocean of air. Middleton Reef bloomed out of the sea like a great blue-and-white orchid. The reef was barely awash and whitely outlined with surf. It was littered with shipwrecks—mostly Japanese fishing boats, thrown up on the table of the reef like discarded toys.

Later we went up to the reef by sea in the steel-hulled fifty-foot fishing boat *Capella*. It was an easy night passage over rare calm seas. The captain, Paul Bowen, told a Middleton Reef shipwreck story. "The last trip up to the reef we found a New Zealand yacht. The Kiwis, two girls and a bloke, kind of flower children, told me about their navigation. They said they 'sailed north from New Zealand for two days, then took a left.' No one told them about Middleton Reef. They were marooned on the reef for three weeks before we found them. The bloke was convinced the girls were going to poison him. The girls were convinced that the bloke was going to murder them. They lived at opposite ends of the yacht, drank rainwater, and ate giant clams from the reef."

At Middleton we anchored near the wreck of the *Runic,* a freezer ship carrying mutton and beef that ran aground in 1961. The reef top was littered with bones. Gary and I dove on a sunken Japanese fishing boat. We found a family of fat black cod living in the galley. One fearlessly adopted me and followed me around for the day, staring jealously over my shoulder as I photographed other fish. The slopes of Middleton Reef were barren; a constantly pounding ocean and a crown-of-thorns starfish infestation had erased much of the live coral. On the second day, a cyclone warning came over the radio. We secured the boat, lashed all the equipment down, and ran south to Lord Howe Island and safety. We could have stayed at Middleton Reef, but might have been stuck for a week or longer. I had a vision of endless meals of giant clams and rainwater.

After a few hours of incessant pounding (the forward bunks of *Capella* actually had seat belts), the twin mountains of Lord Howe Island appeared on the horizon. Six hours later we anchored behind the bulk of the island. The storm increased in fury and for days bent the Kentia palms over and blew through the pillars of Norfolk pines. The outer reefs were a maelstrom of foam and massive breakers. Anne, Emily, and I swam in the protected cove of Ned's Beach, where the islanders had fed fish for decades, producing a school of tame silvery jacks and drummers. It was like swimming in a carousel of fish. An enormous school of stinging catfish lurked around the shadowy pilings beneath the island pier — the largest group of these fish I had ever seen. In the afternoon they moved out across the sandy lagoon floor. They rubbed against each other, twisting and intertwining, while the whole school rolled in slimy nightmare fashion across the sea bed.

One night the tail of the cyclone blew out to sea. In the morning Emily and I wandered down to Lagoon Beach. The swell was whispering against the barrier reef. A dying squid that had washed across the coral from the open ocean glistened a cold midnight blue as its life ended, a peaceful death on a dreamlike island in the far corner of the Pacific.

LORD HOWE
ISLAND,
AUSTRALIA

LAGOON WITH WRASSES AND DAMSELFISH,
LORD HOWE ISLAND, AUSTRALIA

SPLENDID HAWKFISH,
MALABAR REEF,
LORD HOWE ISLAND,
AUSTRALIA

BESPECTACLED ANGELFISH,
THE ADMIRALTIES,
LORD HOWE ISLAND,
AUSTRALIA

153

STINGING CATFISH, LORD HOWE ISLAND PIER, AUSTRALIA

STINGING CATFISH ROLLING ACROSS THE SANDY LAGOON BOTTOM,
LORD HOWE ISLAND, AUSTRALIA

155

BREAKING WAVE, THE ADMIRALTIES,
LORD HOWE ISLAND, AUSTRALIA

MIDDLETON REEF, LORD HOWE RISE,
AUSTRALIA

BLACK COD AND WRECK OF JAPANESE FISHING BOAT,
MIDDLETON REEF, LORD HOWE RISE, AUSTRALIA

158

SPANISH-DANCER NUDIBRANCH IN THE EVENING SUN, MALABAR REEF, LORD HOWE ISLAND, AUSTRALIA

NEW ZEALAND

I am dreaming that I am in the belly of a very large sea creature. The creature is breathing: a slow, rumbling intake, a whooshlike exhale. It is so dark that the darkness has acquired a thickness that I can almost feel. The air is heavy, moist like the air trapped in a lung.

Then I open my eyes. The dream has ended but the breathing sound has not, and the air stays as thick as in my dream.

Three little orange, charging lights from my underwater strobes define the darkness and illuminate the corner of the aft cabin on *Pegasus II*. I climb out of my bunk and go up on deck. The masthead light casts a weak glow over the boat and is reflected in the water. *Pegasus II* rides in its own puddle of light.

We are anchored in a huge sea cave, a cave so vast that it could hold several navy destroyers. The Rikoriko Cave is in the very heart of the Poor Knights Islands, off the North Island of New Zealand. I am at the very end of the Pacific. A westerly swell pushes through the entrance to the back wall and then with a sigh escapes from the cave like the breathing of a giant sea creature at the uttermost part of the sea.

New Zealand is in a wondrous corner of the Pacific, sandwiched between the warm-water belt of Polynesia and the cold Southern Ocean that rings the world with cold, rich waters, waters that support a chain of life that begins with krill (tiny shrimps), goes through penguins, and ends with whales. New Zealand is a land that reminds people of everywhere else — the fiords of Norway, the mountains of Switzerland, the rolling hills of northern England — yet it is New Zealand, a place of clouds and dreams.

PINK JEWEL ANEMONES, POOR KNIGHTS ISLAND, NORTH ISLAND, NEW ZEALAND

BLENNY COURTSHIP, POOR KNIGHTS ISLAND, NORTH ISLAND, NEW ZEALAND

Dawn outlined the rim of the Rikoriko Cave. Eric Gosse, the master of *Pegasus II,* came on deck. Presently we were joined by Dr. Roger Grace, marine biologist, underwater naturalist, and my guide. Silently we looked at the water and the growing light as we slapped the mosquitoes that were beginning their dawn ritual of flying and eating. Eric went into the wheelhouse, took his trombone out of its case, went to the bow of his boat, and blew a very loud, only vaguely melodic chorus of "When the Saints Go Marching In." Below deck in the warm cocoon of *Pegasus II,* someone cursed.

The Poor Knights Islands are off a place called Tutukaka, about midway up the east coast of the North Island. (The name "Tutukaka" brought great joy to my daughter Emily, then six years old, when I talked to her from a red phone box overlooking the harbor.) They say the Poor Knights were named because the islands look like a knight lying on his shield.

Underwater, it is like diving in and around a series of cathedrals, dark volcanic stone cathedrals full of masses of fish. Blue light seeps through entrances, silhouetting, spotlighting, and outlining the inhabitants. The walls of the caves of the Poor Knights are coated with life, jewel anemones of a pink which I had never seen before in nature. Orange-spotted clown nudibranches laid rosettes of eggs and I found two inch-long blennies poking their heads out of the same hole. My huge camera, with its microtelephoto lens, was intruding in the private courtship of the blennies.

But the Poor Knights are only one state in the undersea country of New Zealand. On a clear Sunday morning we took off from Auckland and flew south. Eric piloted a borrowed Cessna 210, an aircraft that likes to pretend it's a miniairliner with its turbo-charged engine. We flew the length of the North Island, then over the water and halfway down the west coast of the South Island, where we were stopped by an impenetrable wall of clouds with huge dark ramparts reaching up to twenty-five thousand feet. Turning around, we went north to the town of Nelson. Though the west coast was socked in, the mountains were clear, and we flew down the spine of the South Island. The Southern Alps of New Zealand are just that: it is as if Switzerland had been put in a vice and squeezed. The air was very rough so we climbed above oxygen level. It was Sunday and we couldn't get the oxygen bottles filled—there was just enough for Eric, who flew while I navigated. Everyone else gasped like fish as New Zealand, from the cloudy Tasman Sea to the brilliant white alps, rolled underneath the wings. Farther south, the alps began to dissolve into a green, pinnacle-studded landscape pressed up against lakes and smooth rolling plains full of sheep. To the west of the pinnacles, the fiords began. Here, it was as if all of the Norwegian coast had been squeezed into a land less than one hundred miles long.

They call this place Fiordland. It has a small perfection to it. Three-thousand-foot mountains fall directly into deep, dark green fiords, which are really arms of the Tasman Sea. One of the entrances to Fiordland is the 2,200-foot Wilmot Pass. I drove across the pass just after an early, unexpected storm dumped over three feet of heavy, slushy Tasman Sea snow. We were following the ranger's snowplow, an ancient British tank tow truck that actually saw service at Al-Alamein. When it got stuck at the top of the pass we transferred our supplies from an equally ancient bus (nothing is discarded in New Zealand) to the park ranger's Land Rovers. We came sliding sideways around a corner and suddenly we saw Doubtful Sound shrouded in cloud, laced by a setting sun, a piece of ocean embraced by the arms of the land.

Fiordland waters are dark green and murky, or so they look from the surface. But the green and the murk are really a layer of fresh water anywhere from two to twenty feet thick, riding on top of clear, cold seawater. A very thin layer of ice coated the surface of a corner of Doubtful Sound, called Crooked Arm. Roger and his diving partner, Linda Ingham, looking like teddy bears bundled in their thick, blue plush New Zealand wet suits, stepped off the boat and into the icy water. I followed. A stab of cold water rolled down my neck as I swam over to the wall of the fiord. Everything looked hazy, out of focus, as if bathed in crystal smoke. I could see the two densities of liquid mixing at the boundary between the fresh- and saltwater layers, tendrils of soft clear clouds entwining. I began to sink and suddenly I was through the freshwater layer. The sea became very clear, green and very dark.

The walls of Crooked Arm are sheer, plunging straight into darkness. Ghostly black coral trees grow out from the wall—a sparse, silent vertical forest. I had come through the looking glass of the surface, gliding into a strange still land. The black coral trees usually grow in the open sea in much deeper water, at two hundred feet or more. They are creatures of low light; the murky freshwater layer acts as a filter. Looking up, I saw Linda swimming above the coral, a black silhouette of a diver. Her exhaust bubbles rose and made holes in the cloudy emerald curtain of the surface.

The walls of Doubtful Sound are covered with large red New Zealand crayfish. The crayfishermen have terrible trouble setting their traps. Occasionally they have to tie them to trees that grow out from the bank.

It rains in Fiordland, a rain of Biblical proportions. Great storms brewed in Antarctica sweep across the Tasman Sea, a far corner of the Pacific, and mash into the green walls of the Fiordland. The rain rolls down the steep green mountains in a thousand waterfalls, and in summer the freshwater layer grows to twenty feet thick. This is a place of instant weather—sun, gray skies, whistling winds, and mirror calms, all in half a day.

On a summer's day when the sea mirrored a gray, cloud-filled sky, a pod of bottle-nosed dolphins came to the bow wave of the research vessel *Renown*. The gray-bodied dolphins slipped in and out of reality, becoming creatures of water and sky. Their home was a winding fiord, a place of sea and mountains.

On another evening, filled with yellow hazy light, I dove beneath a waterfall and found a huge hedgelike black coral tree enveloped in a pink fog. The coral tree was spawning millions of microscopic eggs. Next to the tree I found a group of pink Jason nudibranchs mating and laying eggs. When I came up, the sun had set and the walls and surface of the fiord were bathed in blue twilight, a paradise of waterfalls, of still ocean and dark mountains. When I took my diving hood off, the sandflies attacked. These are the curse of Fiordland, more like flying teeth that not only hurt when they bite but leave a huge bump that itches for days, producing memories you can scratch.

Stewart Island is at the very tip of New Zealand. Here Roger Grace showed me a southern pigfish. The brown, spiky, ten-inch-long creature was resting in the sand at the town harbor. Roger swam over it and did an amazing thing: he picked it up. The creature squirmed slightly as Roger held it gently up to my ear. The pigfish grunted softly. One day, Roger was photographing a pigfish when he discovered that he needed another hand for his camera. He simply dug a hole in the soft sand and "planted" the pigfish, tail first. The animal waited patiently until Roger was photographically finished with it; then it swam out of its parking space, looked calmly back at the camera, and disappeared into the clear green sea.

At the southern end of Stewart Island, we dove with a family of Hooker sea lions. There are only five thousand Hooker sea lions left in the world. They would only play with us between journeys to the open sea, where they consumed mass quantities of squid. They would come into the water at dusk, dark gray dusk, when land and sky would blend together. Underwater it was just possible to see an enormous head, jaws agape and barking, rushing at me. The Hooker sea lions are a perfect combination of giant Labrador retrievers and torpedoes. Some of the creatures were over ten feet long and they would twist and turn, making high-speed runs at us. At one point a seal took my hand in its enormous jaws and daintily, as if teething, nibbled at it.

They are creatures of the boundary between the Pacific and the Southern Ocean. I was at the end of the Pacific. To the north was the great blue belly of the world's greatest ocean; to the south the Southern Ocean, with its roaring winds and rich darkness.

This was the journey's end.

BLUE COD, STEWART ISLAND, NEW ZEALAND

PORTRAIT OF A NEW ZEALAND CRAYFISH, DOUBTFUL SOUND, FIORDLAND, SOUTH ISLAND, NEW ZEALAND

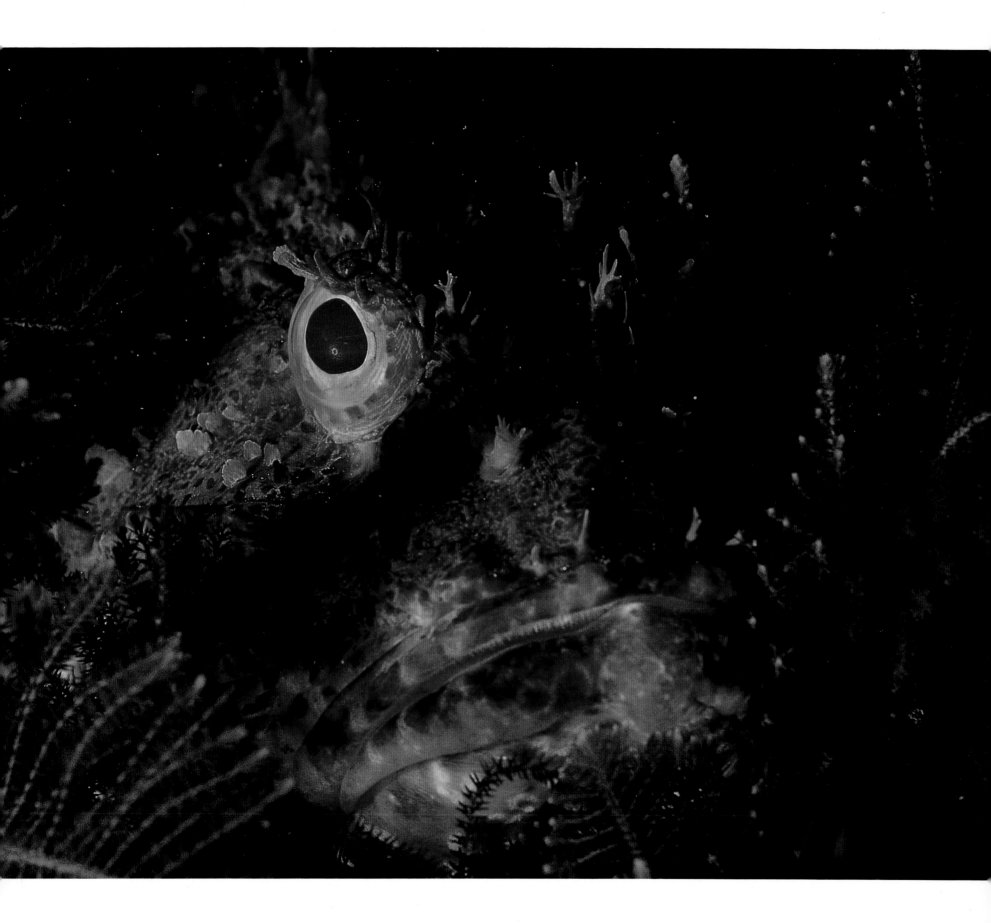

POTBELLIED SEA HORSE, STEWART ISLAND, NEW ZEALAND

SOUTHERN PIGFISH PORTRAIT, POOR KNIGHTS ISLAND, NORTH ISLAND, NEW ZEALAND

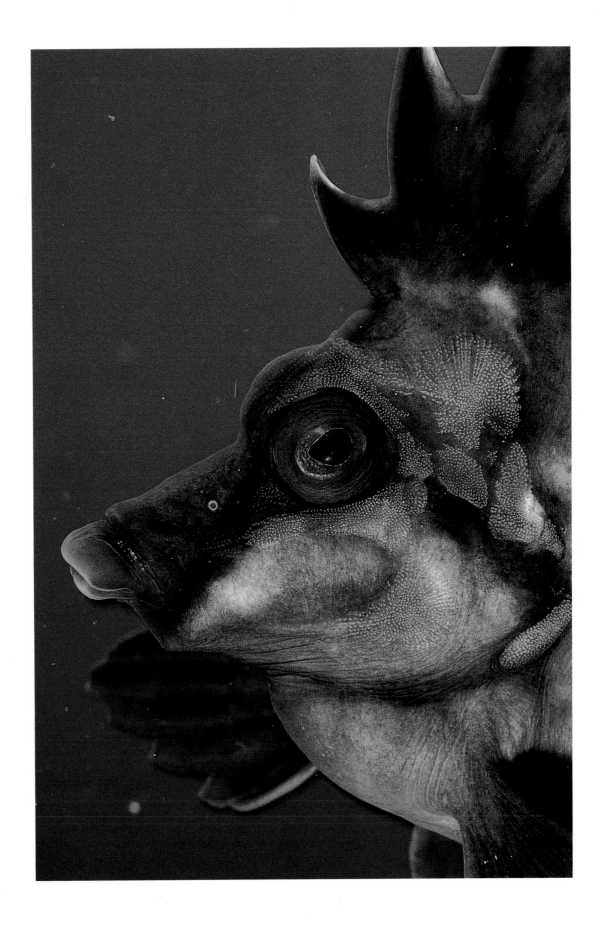

CRESTED TOPKNOT FISH EATING ANOTHER FISH, TOWN PIER,
STEWART ISLAND, NEW ZEALAND

CRESTED TOPKNOT FISH IN SEA LETTUCE, TOWN PIER, STEWART ISLAND, NEW ZEALAND

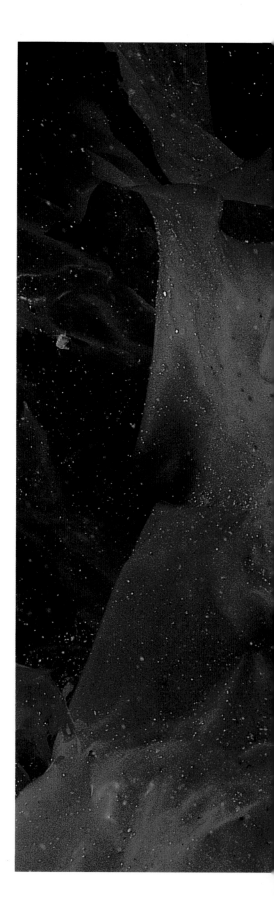

FEMALE HOOKER SEA LION BEING PURSUED BY MALE, STEWART ISLAND, NEW ZEALAND

MOLLYMAWK FEEDS ON FISHING BOAT SCRAPS, STEWART ISLAND, NEW ZEALAND

CROOKED ARM, DOUBTFUL SOUND,
FIORDLAND, SOUTH ISLAND,
NEW ZEALAND

ACKNOWLEDGMENTS

Underwater photography requires an enormous amount of assistance and help. The fish themselves do not really help out. They stare bleakly back at the camera, resenting its intrusion into their secret lives. I would like to thank the following people for their help, encouragement, and patience.

In Monterey: Mark Shelley, Jay Ireland, the Monterey Bay Aquarium Staff, Lynn May, and the crew of the *Silver Prince.*

Off Catalina Island: Doc and Cici White and the crew of the *Mirage.*

In the Strait of Georgia, British Columbia: Doug and Gail Sage, Jim and Jeanie Cosgrove, Warren Buck, Phil Nuytten, and Frank White, Jr.

In Hawaii: Dr. Ricky Grigg, Pete and Jane Cambouras of Central Pacific Divers, Nikolas Konstantinou and Sea Sage Divers, and the Maui Divers Company.

In the Galápagos: Gerard Wellington, the Charles Darwin Research Station, Dr. Fritz Trillmich, Gil Grosvenor, Fiddi Angermeyer, Mike O'Neill, and the crews of the *Beagle III, Bronzewing,* and *Encantada.*

In Palau: Bill and Peggy Hammer.

In Japan: Koji and Miyuki Nakamura, Tadahiko Matsui, Okamura San, Itoh San, Uryu San, Yusuke Yoshino, the crew of *Moby,* Midori and the staff of Osekan, Nikki Konstantinou, and Dai Iwai of *National Geographic.*

In Espíritu Santo: Alan Power and Kevin and Chris Deacon.

In Pearl Harbor: Dan Lineham and the staff of the Submerged Cultural Resource Unit of the National Park Service, the staff of the *Arizona* Memorial, and Daniel Martinez.

In the Solomon Islands: Brian Bailey and family.

In Papua New Guinea: Bob and Dinah Halstead, Bob and Diane Pierce, the crew of the *Telita,* Peter and Henrietta Miller, Steve Birdsell, and Peter and Wendy Benchley.

In the Great Barrier Reef: Rodney and Kay Fox, Andrew Fox, Graham and Elaine McCallum, the crew of the *Auriga Bay,* Phillip and Bev of *Tabasco,* Tracy Aron, Lizzard Island Lodge, Quicksilver Cruises, the staff of Heron Island, and Kenneth Brower.

In Lord Howe Island: Jeff Deacon and family, Peter and Judy Riddle and family, Pat deGroote and Sea Life Diving Services, Paul Bowen and the crew of *Capella,* and Sally and Ted Levine.

In New Zealand: Dr. Roger Grace, Linda Ingham, Eric Gosse and the crew of *Pegasus II,* Lance Shaw and the crew of *Renown,* and Fiordland National Park.

I would like to thank Gary Bell for assistance in the Great Barrier Reef, Lord Howe Island, and New Zealand.

In Los Angeles: Baerwald West.

For the gift of time in the sea, the golden opportunity of an assignment, I would like to thank former Editor-in-Chief of *National Geographic* Bill Garrett and the present Editor, Bill Graves. Bill Graves is also responsible for

43RD INFANTRY DIVISION'S TYPEWRITER, *PRESIDENT COOLIDGE,* PAPUA NEW GUINEA

my writing career. He uses the time-honored carrot-and-stick approach, sometimes more carrot. Sometimes more stick.

At *Geographic* I would like to thank Tom Kennedy, Kent Kobersteen, and former Directors of Photography Bob Gilka and Rich Clarkson for their support.

The Suruga Bay, Lord Howe Island, New Zealand, and the *President Coolidge* sections grew out of stories I did for *National Geographic.* For putting up with my literary excesses, I would like to thank the superb editors at *Geographic:* Pritt Vesilind, Don Belt, Jane Vessells, David Jeffery, and Bart McDowell.

On the picture side, I would like to thank David Arnold, Bob Patton, Mary Smith, and Bill Allen.

My thanks to Al Royce and Marissa Domeyko for all their help. I would like to thank Bob Madden, Connie Phelps, Kate Glassner, and David Griffin. I could not have

EQUIPMENT

gotten there and gotten back without the help of the National Geographic Travel Division.

I would like to thank Bethany Judt and Charlene Valeri.

Jennifer Angle deciphered my underwater cuniform (a written language where bad handwriting disguises bad spelling) and edited and translated it into a manuscript.

I would like to thank Janet Bush and Patricia Hansen for their patience and wonderful editing, Amanda Freymann for overseeing the production of the book, and Susan Marsh for the book's design. My thanks also to my agent, Robin Straus.

I would especially like to thank Dr. Eugenie Clark, who constantly opens my eyes to the sea.

In Boulder, Colorado: Ann Doubilet.
In Elberon, New Jersey: Bobby Doubilet.
I would like to thank my sister Jane Doubilet Kramer and my niece Raine for braving the mighty deep.

And finally, Anne and Emily Doubilet, together we have journeyed across this largest of oceans and seen the beauty of its silent lands.

David Doubilet
Elberon, New Jersey
February 1992

I usually go on assignment with seven or more underwater housings and several Nikonos V cameras with 15mm lenses. Underwater it is impossible to change lenses or film, so to do the same job a normal photojournalist does with two camera bodies and a half a dozen lenses I need a vast mountain of equipment. Sometimes I travel with as many as twenty cases of camera gear and frankly, it's killing me. Here is what I use as my basic kit: Nikon F-3s with DA-2 Action Finders and MD-4 motor drives and cast-aluminum AquaVision Aquatica IIIN housings. For lenses I use Nikkor 13mm, 15mm, 16mm fish-eye, 18mm, 20mm, 24mm, 28mm, 55mm micro, and 105mm micro. For close-up work, I use 60 and 100mm Zeiss macro Planar lenses, originally made for Contax cameras; Joe Stancampiano has converted these lenses to work on Canon F-1 cameras, which I put in the Aquatica IIIC housings. The F-1 cameras have sports finders and motor drives. I also use the Canon 200mm micro lens in a modified Aquatica IIIC housing. I use Sea and Sea YS 200 strobes and they have adaptors that allow me to use them with E. O. Bulkhead connectors.

I also use Sonic Research SR2000 slave units. Occasionally I will use Subatec 180 and 300 movie lights for constant-burning supplemental lighting.

I use Kodachrome 25, 64, and 200 ASA films. They are fine-grained, long-lasting, and do not produce a fake blue.

None of the pictures in this book would have been possible without the support and genius of the *National Geographic*'s Photo Equipment Shop. Joe Stancampiano makes ideas into cameras. Kenji Yamaguchi maintains the cameras and has produced elegant, simple ways to make new images. Keith Moorehead keeps the housings afloat and also takes my wandering daydreams and turns them into aluminum and plexiglass reality.

The strobes are maintained by the strobe king, Larry Kinney. Phil Leonardi works on the electronics. Sergio Ballivian, Mary Simone, and Teresa Corrales are the support staff. Nelson Brown is the director of the shop.

Underwater photography is a strange combination of hunting, swimming, ocean knowledge, and visual philosophy. I depend for my life and my vision on machines that deliver technical perfection, day in, day out. I may see the fish and push the trigger, but it is a collaborative effort. The people of the Photo Equipment Shop are with me all the time in the sea.

Any underwater photographer who says "I don't care about equipment, I just care about the images" will not make pictures. The overly philosophical undersea photographer will probably end up being swallowed, when not paying attention, by some large sea creature.

LATIN NAMES

BLENNY, THE BOMMIE, HERON ISLAND, CAPRICORN GROUP, GREAT BARRIER REEF, AUSTRALIA

Library of Congress Cataloging-in-Publication Data

Doubilet, David.

   Pacific: an undersea journey / by David Doubilet. — 1st ed.

      p.   cm.

   ISBN 0-8212-1903-0

   1. Marine biology — Pacific Ocean — Pictorial works.   2. Underwater photography.   1. Title.

QH95.D68 1992

574.92'5'0222 — dc20   92-4514

RED GORGONIAN CORAL WITH FEEDING POLYPS,

FUTO POINT, IZU PENINSULA, JAPAN

DESIGNED BY SUSAN MARSH

TYPE SET IN GILL SANS BY MONOTYPE COMPOSITION COMPANY

PRINTED AND BOUND BY DAI NIPPON PRINTING COMPANY, LTD.